Theory of
Electromagnetoelasticity

Theory of
Electromagnetoelasticity

Jiashi Yang

University of Nebraska-Lincoln, USA

NEW JERSEY · LONDON · SINGAPORE · BEIJING · SHANGHAI · HONG KONG · TAIPEI · CHENNAI · TOKYO

Published by

World Scientific Publishing Co. Pte. Ltd.
5 Toh Tuck Link, Singapore 596224
USA office: 27 Warren Street, Suite 401-402, Hackensack, NJ 07601
UK office: 57 Shelton Street, Covent Garden, London WC2H 9HE

Library of Congress Cataloging-in-Publication Data
Names: Yang, Jiashi, 1956– author.
Title: Theory of electromagnetoelasticity / Jiashi Yang, University of Nebraska-Lincoln, USA.
Description: New Jersey : World Scientific, [2024] | Includes bibliographical references and index.
Identifiers: LCCN 2023032349 | ISBN 9789811281884 (hardcover) |
     ISBN 9789811281891 (ebook for institutions) | ISBN 9789811281907 (ebook for individuals)
Subjects: LCSH: Piezoelectric materials. | Electromagnetic theory. | Elasticity.
Classification: LCC QC595.5 .Y363 2024 | DDC 537/.2446--dc23/eng/20231120
LC record available at https://lccn.loc.gov/2023032349

British Library Cataloguing-in-Publication Data
A catalogue record for this book is available from the British Library.

For any available supplementary material, please visit
https://www.worldscientific.com/worldscibooks/10.1142/13557#t=suppl

Desk Editors: Logeshwaran Arumugam/Amanda Yun

Typeset by Stallion Press
Email: enquiries@stallionpress.com

Printed in Singapore

To the memory of
my mother Huiyun Wang (王惠云)
and my father
Shih-te Yang (杨式德)

Preface

In the conventional theory of piezoelectricity, the electric field is quasistatic. As a consequence the theory cannot describe electromagnetic waves and acousto-optic interactions. The quasistatic theory of magnetoelasticity is similar. To model acoustic, electric, magnetic and optic interactions, the quasistatic approximation has to be abandoned and a fully dynamic theory with electric and magnetic fields governed by Maxwell's equations is needed. This book presents the construction of such a dynamic theory for the electrodynamics of elastic solids.

The first three chapters are on uncoupled theories of electromagnetism and elasticity. They serve as a preparation for later chapters for coupled electromagnetoelastic fields which are more involved. Specifically, Chapter 1 is on classical electrodynamics which has electromagnetic coupling but does not have couplings to acoustic fields. Chapter 2 is on continuum kinematics which along with Chapter 3 form the theory of elasticity.

The continuum theory of the electrodynamics of elastic solids is developed in Chapters 4–6. The standard approach of constructing the theory of continuum electrodynamics is based on the electrodynamics of charged particles followed by the transition to a continuum. However, this book employs a model of two charged and interacting continua established by H.F. Tiersten. The two-continuum model is presented in Chapter 4, followed by balance laws in Chapter 5 and constitutive relations in Chapter 6.

The theory developed in Chapters 4–6 is complicated due to various nonlinearities associated with large deformations and strong fields, and presents considerable mathematical challenges. For the special case of the linear theory for small deformations and weak fields, a series of reasonably simple solutions are given in Chapter 7 for polarizable but nonmagnetizable materials.

In Chapter 8, the nonlinear theory is extended to include thermal couplings and a few dissipative effects such as electrical conduction.

The literature on electromagnetoelasticity are numerous. Only those books or papers whose results are used directly in the present book are listed as references. No attempt was made to review the literature. Because of the multi-physical fields involved, some symbols may have different meanings in different places of the book. For convenience, a list of symbols is given in Appendix A. A table for the basic equations of electrodynamics in SI and Gaussian units is provided in Appendix B. Appendix C gathers a few vector identities used in various places of the book. Some common material constants are collected in Appendix D.

About the Author

 **Jiashi Yang** is a Full Professor at the Department of Mechanical and Materials Engineering of University of Nebraska-Lincoln. He received his bachelor's and master's degrees from Tsinghua University in 1982 and 1985, and his Ph.D. from Princeton University in 1994. Then he did postdoctoral research at University of Missouri-Rolla and Rensselaer Polytechnic Institute. He worked as an engineer with Motorola, Inc. prior to joining University of Nebraska in 1997. His research area is the mechanics of electromagnetoelastic structures and devices. His previous books include *Mechanics of Piezoelectric Structures* and *Mechanics of Functional Materials* with World Scientific, as well as *An Introduction to the Theory of Piezoelectricity* and *Analysis of Piezoelectric Semiconductor Structures* with Springer.

Contents

Chapter 1

Electromagnetism

This chapter presents the basics of classical electromagnetism [1–4]. They are about electric and magnetic fields in a vacuum or rigid materials without deformation. The materials are stationary except in the last section. In this chapter, the spatial coordinates are x_k or $\mathbf{x}$. The Cartesian tensor notation is used, along with the summation convention for repeated tensor indices and the convention that a comma followed by an index denotes partial differentiation with respect to the coordinate associated with the index. V, S and C represent volumes, surfaces and curves fixed in space.

1.1 Electrostatics in Vacuum

According to Coulomb's law between two point charges, Q and Q', the force $\mathbf{F}$ and electric field $\mathbf{E}$ on Q' at a position $\mathbf{r} = \mathbf{x}$ from Q (see Fig. 1.1) are given by

$$
\mathbf{F} = \frac{QQ'}{4\pi\varepsilon_0 r^3}\mathbf{r} = Q'\mathbf{E},
$$

$$
\mathbf{E}(\mathbf{x}) = \frac{Q}{4\pi\varepsilon_0}\frac{\mathbf{r}}{r^3} = \frac{Q}{4\pi\varepsilon_0}\nabla\left(\frac{-1}{r}\right),
$$

(1.1.1)

where ε_0 is the electric permittivity of free space, and we have used

$$
\frac{\mathbf{r}}{r^3} = \nabla\left(\frac{-1}{r}\right), \quad \nabla = \mathbf{e}_i\frac{\partial}{\partial x_i}.
$$

(1.1.2)

$\mathbf{e}_i$ are the unit vectors along x_i.

1

Fig. 1.1.   Two point charges.

From Eq. (1.1.1), for a closed surface S enclosing Q,

$$\int_S \mathbf{E} \cdot \mathbf{dS} = \int_S \frac{Q}{4\pi\varepsilon_0 r^2} \frac{\mathbf{r}}{r} \cdot \mathbf{dS}$$

$$= \int_S \frac{Q}{4\pi\varepsilon_0} d\Omega = \frac{Q}{4\pi\varepsilon_0} \int_S d\Omega = \frac{Q}{4\pi\varepsilon_0} 4\pi = \frac{Q}{\varepsilon_0}, \quad (1.1.3)$$

where

$$\frac{1}{r^2} \frac{\mathbf{r}}{r} \cdot \mathbf{dS} = d\Omega \qquad (1.1.4)$$

has been used. $d\Omega$ is the solid angle corresponding to $\mathbf{dS}$. For a closed surface, the solid angle is 4π. The divergence of the electric field $\mathbf{E}$ in Eq. (1.1.1) is given by

$$\nabla \cdot \mathbf{E} = \frac{Q}{\varepsilon_0} \nabla \cdot \left(\frac{\mathbf{r}}{4\pi r^3} \right) = \frac{Q}{\varepsilon_0} \delta(\mathbf{r}), \qquad (1.1.5)$$

where

$$\delta(\mathbf{r}) = \frac{1}{4\pi} \nabla \cdot \left(\frac{\mathbf{r}}{r^3} \right) = \frac{1}{4\pi} \nabla \cdot \nabla \left(\frac{-1}{r} \right)$$

$$= \frac{1}{4\pi} \nabla^2 \left(\frac{-1}{r} \right),$$

$$\nabla^2 = \nabla \cdot \nabla. \qquad (1.1.6)$$

δ is the Dirac delta function. Mathematically, Eq. $(1.1.6)_1$ shows that $-1/(4\pi r)$ is the so-called fundamental solution of the Laplace operator. By superposition, in the case of a continuous distribution of charges with density ρ^T per unit volume occupying a region V, we

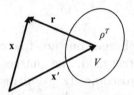

Fig. 1.2. A continuous distribution of charges in a region V.

obtain

$$\mathbf{E}(\mathbf{x}) = \int_V \frac{\rho^T(\mathbf{x}')}{4\pi\varepsilon_0 r^2} \frac{\mathbf{r}}{r} dV', \tag{1.1.7}$$

$$\nabla \cdot \mathbf{E} = \frac{\rho^T}{\varepsilon_0}, \tag{1.1.8}$$

where $\mathbf{r} = \mathbf{x} - \mathbf{x}'$ as shown in Fig. 1.2, $\mathbf{x}' \in V$, and the operator ∇ is with respect to $\mathbf{x}$.

From Eq. (1.1.1), for a closed curve C, we have

$$\oint_C \mathbf{E} \cdot \mathbf{dl} = \oint_C \frac{Q}{4\pi\varepsilon_0 r^2} \frac{\mathbf{r}}{r} \cdot \mathbf{dl}$$

$$= \oint_C \frac{Q}{4\pi\varepsilon_0} \frac{1}{r^2} dr = -\oint_C \frac{Q}{4\pi\varepsilon_0} d\left(\frac{1}{r}\right) = 0, \tag{1.1.9}$$

where $\mathbf{dl}$ is the differential line element along the curve and

$$\frac{\mathbf{r}}{r} \cdot \mathbf{dl} = dr \tag{1.1.10}$$

has been used. Equation (1.1.9) implies, through Stokes' theorem,

$$\nabla \times \mathbf{E} = 0. \tag{1.1.11}$$

Then an electrostatic potential φ can be introduced such that

$$\mathbf{E} = -\nabla\varphi. \tag{1.1.12}$$

The substitution of Eq. (1.1.12) into Eq. (1.1.8) results in a single equation for φ, i.e.,

$$\nabla \cdot \mathbf{E} = -\nabla \cdot (\nabla\varphi) = -\nabla^2\varphi = \frac{\rho^T}{\varepsilon_0}. \tag{1.1.13}$$

1.2 Dielectrics

Dielectrics have bound charges but not free charges. When a dielectric is placed in an electric field, the charges in its molecules redistribute themselves microscopically, resulting in a macroscopically polarized state. The microscopic charge redistribution may occur in different ways (see Fig. 1.3).

At the macroscopic level, the distinctions among different polarization mechanisms do not matter. A macroscopic polarization vector per unit volume,

$$\mathbf{P} = \lim_{\Delta V \to 0} \frac{1}{\Delta V} \sum_{\Delta V} \mathbf{p}, \tag{1.2.1}$$

is introduced which describes the macroscopic polarized state of the material. For a dielectric body occupying a region V with a boundary surface S, the effective polarization charge densities are [5]

$$\rho^p = -P_{i,i} = -\nabla \cdot \mathbf{P} \quad \text{in} \quad V,$$
$$\sigma^p = n_i P_i = \mathbf{n} \cdot \mathbf{P} \quad \text{on} \quad S, \tag{1.2.2}$$

where $\mathbf{n}$ is the outward unit normal of S. The total effective polarization charge of the body is given by

$$\int_V \rho^p dV + \int_S \sigma^p dS$$
$$= \int_V -P_{k,k} dV + \int_S n_k P_k dS$$
$$= \int_V -P_{k,k} dV + \int_V P_{k,k} dV = 0, \tag{1.2.3}$$

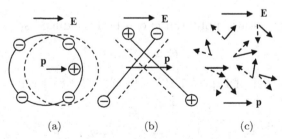

Fig. 1.3. Microscopic polarizations. (a) Electronic, (b) ionic, and (c) orientational.

which is as expected. The total electric moment produced by the effective polarization charges is

$$\int_V x_j \rho^P dV + \int_S x_j \sigma^P dS$$

$$= \int_V x_j(-P_{k,k})dV + \int_S x_j n_k P_k dS$$

$$= \int_V x_j(-P_{k,k})dV + \int_V (x_j P_k)_{,k} dV = \int_V P_j dV, \quad (1.2.4)$$

which is also as expected.

With the effective polarization charges, the charge equation of electrostatics in Eq. (1.1.8) may be written as

$$\nabla \cdot \mathbf{E} = E_{k,k} = \frac{\rho^T}{\varepsilon_0} = \frac{1}{\varepsilon_0}(\rho^P + \rho^E)$$

$$= \frac{1}{\varepsilon_0}(-P_{k,k} + \rho^E), \quad (1.2.5)$$

or

$$(\varepsilon_0 E_k + P_k)_{,k} = \rho^E, \quad (1.2.6)$$

where ρ^E represents charges other than the effective polarization charges. It is usually zero in a dielectric and is kept formally only. With the introduction of an electric displacement vector $\mathbf{D}$ by

$$D_k = \varepsilon_0 E_k + P_k, \quad (1.2.7)$$

Eq. (1.2.6) can be written as

$$\nabla \cdot \mathbf{D} = D_{k,k} = \rho^E. \quad (1.2.8)$$

Using the effective volume and surface polarization charges, the electric body force $\mathbf{f}^E$ due to polarization can be calculated from

$$\int_V \rho^P E_j dV + \int_S \sigma^P E_j dS$$

$$= \int_V -P_{k,k} E_j dV + \int_S n_k P_k E_j dS$$

$$= \int_V -P_{k,k} E_j dV + \int_V (P_k E_j)_{,k} dS = \int_V f_j^E dV, \quad (1.2.9)$$

where

$$f_j^E = P_k E_{j,k}, \quad \mathbf{f}^E = \mathbf{P} \cdot \nabla \mathbf{E}. \tag{1.2.10}$$

It will be proven convenient to introduce an electrostatic stress tensor $\mathbf{T}^E$ [6] whose divergence yields the electric body force $\mathbf{F}^E = \rho^E \mathbf{E} + \mathbf{f}^E$ through

$$T_{ij,i}^E = \rho^E E_j + f_j^E = F_j^E. \tag{1.2.11}$$

For the existence of such a $\mathbf{T}^E$, consider

$$T_{ij}^E = D_i E_j - \frac{1}{2} \varepsilon_0 E_k E_k \delta_{ij}. \tag{1.2.12}$$

We have

$$T_{ij,i}^E = D_{i,i} E_j + (\varepsilon_0 E_i + P_i) E_{j,i} - \varepsilon_0 E_k E_{k,j}$$
$$= \rho^E E_j + P_i E_{j,i} = \rho^E E_j + f_j^E = F_j^E. \tag{1.2.13}$$

We also have

$$\varepsilon_{ijk} T_{jk}^E = \varepsilon_{ijk} P_j E_k. \tag{1.2.14}$$

Obviously,

$$\int_V (\rho^E E_i + f_i^E) dV = \int_V T_{ji,j}^E dV = \int_S n_j T_{ji}^E dS. \tag{1.2.15}$$

We note that the $\mathbf{T}^E$ as defined by Eq. (1.2.11) is not unique in the sense that there are other tensors that also satisfy Eq. (1.2.11). For example, adding a second-rank tensor with a zero divergence to the $\mathbf{T}^E$ in Eq. (1.2.12) does not affect its satisfaction of Eq. (1.2.11).

The electric body couple $\mathbf{C}^E$ due to polarization can be calculated from

$$\int_V \varepsilon_{ijk} x_j \rho^p E_k dV + \int_S \varepsilon_{ijk} x_j \sigma^p E_k dS$$

$$= \int_V \varepsilon_{ijk} x_j (-P_{m,m}) E_k dV + \int_S \varepsilon_{ijk} x_j n_m P_m E_k dS$$

$$= \int_V [C_i^E + \varepsilon_{ijk} x_j f_k^E] dV, \qquad (1.2.16)$$

where

$$C_i^E = \varepsilon_{ijk} P_j E_k = \varepsilon_{ijk} T_{jk}^E. \qquad (1.2.17)$$

For a linear material,

$$P_k = \varepsilon_0 \chi_{kl}^E E_l, \qquad (1.2.18)$$

where χ_{kl}^E is the electric susceptibility tensor. Then

$$D_k = \varepsilon_0 E_k + P_k = \varepsilon_0 \delta_{kl} E_l + \varepsilon_0 \chi_{kl}^E E_l$$

$$= (\varepsilon_0 \delta_{kl} + \varepsilon_0 \chi_{kl}^E) E_l = \varepsilon_{kl} E_l, \qquad (1.2.19)$$

where

$$\varepsilon_{kl} = \varepsilon_0 (\delta_{kl} + \chi_{kl}^E). \qquad (1.2.20)$$

ε_{kl} is the electric permittivity tensor. In matrix notation,

$$\begin{bmatrix} D_1 \\ D_2 \\ D_3 \end{bmatrix} = \begin{bmatrix} \varepsilon_{11} & \varepsilon_{12} & \varepsilon_{13} \\ \varepsilon_{21} & \varepsilon_{22} & \varepsilon_{23} \\ \varepsilon_{31} & \varepsilon_{32} & \varepsilon_{33} \end{bmatrix} \begin{bmatrix} E_1 \\ E_2 \\ E_3 \end{bmatrix}. \qquad (1.2.21)$$

The energy density of a dielectric is (see Eq. (1.10.11))

$$\hat{U}(\mathbf{D}) = \frac{1}{2} E_i D_i. \qquad (1.2.22)$$

An electric enthalpy density H^E can be introduced through the following Legendre transform:

$$H^E(\mathbf{E}) = \hat{U}(\mathbf{D}) - E_i D_i = -\frac{1}{2} \varepsilon_{ij} E_i E_j, \qquad (1.2.23)$$

which shows that only the symmetric part of ε_{kl} matters. Then Eq. (1.2.21) can be calculated from

$$D_i = -\frac{\partial H^E}{\partial E_i} = \varepsilon_{ik}E_k. \tag{1.2.24}$$

The electric field-potential relation is

$$\mathbf{E} = -\nabla\varphi, \quad E_i = -\varphi_{,i}. \tag{1.2.25}$$

Then

$$D_k = \varepsilon_{kl}E_l = -\varepsilon_{kl}\varphi_{,l}. \tag{1.2.26}$$

The substitution of Eq. (1.2.26) into Eq. (1.2.8) leads to a single equation for φ:

$$-(\varepsilon_{kl}\varphi_{,l})_{,k} = \rho^E. \tag{1.2.27}$$

1.3 Conductors

In conductors there are positive charges fixed to the lattice, bound charges responsible for polarization, and free electrons that can flow through the lattice. Consider the case when a conductor is homogeneous and electrically neutral in a reference state without fields, currents and net charges. Under a voltage or electric field, the free electrons move to form currents and charge distributions. The electric field-potential relation and the charge equation of electrostatics are the same as those in Section 1.2:

$$E_k = -\varphi_{,k}, \quad D_k = \varepsilon_{kl}E_l, \tag{1.3.1}$$

$$D_{k,k} = \rho^E. \tag{1.3.2}$$

For a linear conductor, the current density $\mathbf{J}$ is proportional to the electric field $\mathbf{E}$ through Ohm's law

$$J_k = \sigma_{kl}E_l, \tag{1.3.3}$$

or

$$\begin{bmatrix} J_1 \\ J_2 \\ J_3 \end{bmatrix} = \begin{bmatrix} \sigma_{11} & \sigma_{12} & \sigma_{13} \\ \sigma_{21} & \sigma_{22} & \sigma_{23} \\ \sigma_{31} & \sigma_{32} & \sigma_{33} \end{bmatrix} \begin{bmatrix} E_1 \\ E_2 \\ E_3 \end{bmatrix}, \tag{1.3.4}$$

where σ_{kl} is the conductivity and

$$\sigma_{kl} = \sigma_{lk}. \tag{1.3.5}$$

In terms of the electric potential φ,

$$J_k = \sigma_{kl} E_l = -\sigma_{kl}\varphi_{,l}. \tag{1.3.6}$$

The conservation of charge or continuity equation takes the following form (see Eq. (1.9.9)):

$$\frac{\partial \rho^E}{\partial t} = -J_{k,k}, \tag{1.3.7}$$

which, with the use of Eq. (1.3.6), becomes

$$\frac{\partial \rho^E}{\partial t} = (\sigma_{kl}\varphi_{,l})_{,k}. \tag{1.3.8}$$

The charge equation of electrostatics in Eq. (1.3.2) can be written as

$$D_{k,k} = -(\varepsilon_{kl}\varphi_{,l})_{,k} = \rho^E. \tag{1.3.9}$$

Equations (1.3.8) and (1.3.9) are two equations for ρ^E and φ.

In the special case of an isotropic conductor, we have

$$\varepsilon_{kl} = \varepsilon\delta_{lk}, \quad \sigma_{kl} = \sigma\delta_{lk}. \tag{1.3.10}$$

Then

$$\varepsilon\nabla \cdot \mathbf{E} = \rho^E,$$
$$\dot{\rho}^E = -\nabla \cdot \mathbf{J}, \tag{1.3.11}$$
$$\mathbf{J} = \sigma\mathbf{E},$$

where, for simplicity, we have used a superimposed dot for the time derivative with $\mathbf{x}$ fixed in the case of rigid and stationary materials in this chapter. With substitutions from Eqs. $(1.3.11)_{1,3}$, we can write

Eq. $(1.3.11)_2$ as

$$\dot{\rho}^E = -\frac{\sigma}{\varepsilon}\rho^E. \tag{1.3.12}$$

Equation (1.3.12) can be integrated to produce

$$\rho^E(t) = \rho^E(0)\exp\left(-\frac{\sigma}{\varepsilon}t\right)$$

$$= \rho^E(0)\exp\left(-\frac{t}{\tau}\right), \tag{1.3.13}$$

where

$$\tau = \frac{\varepsilon}{\sigma} \tag{1.3.14}$$

is the so-called relaxation time of a conductor which describes the time needed to reach an essentially static state after an initial disturbance.

1.4   Semiconductors

In semiconductors, in addition to the effective polarization charges, there are charges from doping which are imbedded in the lattice and mobile charge carriers of holes and electrons which are responsible for semiconduction. We assume that in the reference state the material is uniform and electrically neutral. The equations of electrostatics are

$$E_k = -\varphi_{,k},$$

$$D_k = \varepsilon_{kl}E_l, \tag{1.4.1}$$

$$D_{k,k} = q(p - n + N_D^+ - N_A^-),$$

where q is the elementary charge. p and n are the concentrations of holes and electrons. N_A^- and N_D^+ are the concentrations of ionized acceptors and donors from doping. N_A^- and N_D^+ produce holes and electrons which contribute to p and n, respectively. The continuity equations for the holes and electrons are

$$qp = -J^p_{i,i} + \gamma^p,$$
$$qn = J^n_{i,i} + \gamma^n,$$

(1.4.2)

where $\mathbf{J}^p$ and $\mathbf{J}^n$ are the hole and electron current densities. γ^p and γ^n are the sources of holes and electrons. They may be from mechanical, thermal, electrical, magnetic and optical origins. For a macroscopic theory, specific expressions of γ^p and γ^n belong to the so-called constitutive relations. They can be determined experimentally or from microscopic theories. The constitutive relations for the current densities are

$$J^p_i = qp\mu^p_{ij}E_j - qD^p_{ij}p_{,j},$$
$$J^n_i = qn\mu^n_{ij}E_j + qD^n_{ij}n_{,j},$$

(1.4.3)

where $\boldsymbol{\mu}^p$ and $\boldsymbol{\mu}^n$ are the mobility tensors of holes and electrons, respectively. $\mathbf{D}^p$ and $\mathbf{D}^n$ are the diffusion constants. The first term in $\mathbf{J}^p$ (or $\mathbf{J}^n$) is the drift current which is nonlinear as a product of the carrier concentration p (or n) and the electric field $\mathbf{E}$. The second term in $\mathbf{J}^p$ (or $\mathbf{J}^n$) is the diffusion current. The mobility and diffusion constants satisfy the Einstein relation, e.g.,

$$\frac{\mu^p_{33}}{D^p_{33}} = \frac{\mu^n_{33}}{D^n_{33}} = \frac{q}{k_B\theta},$$

(1.4.4)

where k_B is the Boltzmann constant and θ the absolute temperature. With substitutions from Eqs. $(1.4.1)_{1,2}$ and Eq. (1.4.3), we can write Eqs. $(1.4.1)_3$ and (1.4.2) as three equations for φ, p and n.

1.5 Magnetostatics in Vacuum

Consider a current density distribution $\mathbf{J}^T$ in a region V. According to the Biot–Savart law, the magnetic induction $\mathbf{B}$ and its force $\mathbf{F}$ on a current element $I\mathbf{dl}$ at a point $\mathbf{x}$ is given by

$$\mathbf{B}(\mathbf{x}) = \frac{\mu_0}{4\pi} \int_V \mathbf{J}^T(\mathbf{x}') \times \frac{\mathbf{r}}{r^3} dV'$$

$$= \frac{\mu_0}{4\pi} \int_V \mathbf{J}^T(\mathbf{x}') \times \nabla \left(\frac{-1}{r}\right) dV', \qquad (1.5.1)$$

$$d\mathbf{F} = I\mathbf{dl} \times \mathbf{B},$$

where $\mathbf{x}' \in V, \mathbf{r} = \mathbf{x} - \mathbf{x}'$ (see Fig. 1.2 with ρ^T replaced by $\mathbf{J}^T$), and Eq. (1.1.2) has been used. The operator ∇ is with respect to $\mathbf{x}$. μ_0 is the magnetic permeability of free-space. Then

$$\nabla \cdot \mathbf{B} = \frac{\mu_0}{4\pi} \int_V \nabla \cdot \left\{ \mathbf{J}^T(\mathbf{x}') \times \nabla \left(\frac{-1}{r}\right) \right\} dV'$$

$$= \frac{\mu_0}{4\pi} \int_V \left\{ [\nabla \times \mathbf{J}^T(\mathbf{x}')] \cdot \frac{\mathbf{r}}{r^3} - \mathbf{J}^T(\mathbf{x}') \cdot \left[\nabla \times \nabla \left(\frac{-1}{r}\right) \right] \right\} dV'$$

$$= \frac{\mu_0}{4\pi} \int_V \{\mathbf{0} - \mathbf{0}\} dV' = 0, \qquad (1.5.2)$$

where the following vector identity [7] has been used:

$$\nabla \cdot (\mathbf{a} \times \mathbf{b}) = (\nabla \times \mathbf{a}) \cdot \mathbf{b} - \mathbf{a} \cdot (\nabla \times \mathbf{b}). \qquad (1.5.3)$$

For a scalar field f and a vector field $\mathbf{a}$, we have [7]

$$\nabla \times (f\mathbf{a}) = (\nabla f) \times \mathbf{a} + f(\nabla \times \mathbf{a}),$$

$$\nabla \cdot (f\mathbf{a}) = (\nabla f) \cdot \mathbf{a} + f(\nabla \cdot \mathbf{a}). \qquad (1.5.4)$$

According to Eq. $(1.5.4)_1$,

$$\nabla \times \left[\frac{1}{r} \mathbf{J}^T(\mathbf{x}') \right] = \left(\nabla \frac{1}{r} \right) \times \mathbf{J}^T(\mathbf{x}') + \frac{1}{r} \nabla \times \mathbf{J}^T(\mathbf{x}')$$

$$= \left(\nabla \frac{1}{r} \right) \times \mathbf{J}^T(\mathbf{x}') + \mathbf{0} = \left(\nabla \frac{1}{r} \right) \times \mathbf{J}^T(\mathbf{x}').$$

$$(1.5.5)$$

Then Eq. $(1.5.1)_1$ can be written as

$$\mathbf{B}(\mathbf{x}) = \frac{\mu_0}{4\pi} \int_V \nabla \left(\frac{1}{r} \right) \times \mathbf{J}^T(\mathbf{x}') dV'$$

$$= \frac{\mu_0}{4\pi} \int_V \nabla \times \left[\frac{1}{r} \mathbf{J}^T(\mathbf{x}') \right] dV'$$

$$= \frac{\mu_0}{4\pi} \nabla \times \int_V \frac{\mathbf{J}^T(\mathbf{x}')}{r} dV' = \nabla \times \mathbf{A}, \qquad (1.5.6)$$

where we have denoted

$$\mathbf{A} = \frac{\mu_0}{4\pi} \int_V \frac{\mathbf{J}^T(\mathbf{x}')}{r} dV'. \qquad (1.5.7)$$

$\mathbf{A}$ is the vector potential of $\mathbf{B}$. Then the divergence of $\mathbf{B}$ can also be calculated from

$$\nabla \cdot \mathbf{B} = \nabla \cdot (\nabla \times \mathbf{A}) = 0. \qquad (1.5.8)$$

With $\mathbf{A}$, by another vector identity [7], the curl of $\mathbf{B}$ can be written as

$$\nabla \times \mathbf{B} = \nabla \times (\nabla \times \mathbf{A}) = \nabla(\nabla \cdot \mathbf{A}) - \nabla^2 \mathbf{A}. \qquad (1.5.9)$$

For $\mathbf{A}$, we have

$$\nabla \cdot \mathbf{A} = \frac{\mu_0}{4\pi} \int_V \nabla \cdot \left[\frac{\mathbf{J}^T(\mathbf{x}')}{r} \right] dV'$$

$$= \frac{\mu_0}{4\pi} \int_V \left[\nabla \left(\frac{1}{r} \right) \cdot \mathbf{J}^T(\mathbf{x}') + \frac{1}{r} \nabla \cdot \mathbf{J}^T(\mathbf{x}') \right] dV'$$

$$= \frac{\mu_0}{4\pi} \int_V \left[-\nabla' \left(\frac{1}{r} \right) \cdot \mathbf{J}^T(\mathbf{x}') + 0 \right] dV'$$

$$= -\frac{\mu_0}{4\pi} \int_V \nabla' \cdot \left[\mathbf{J}^T(\mathbf{x}') \frac{1}{r} \right] dV' + \frac{\mu_0}{4\pi} \int_V \frac{1}{r} \nabla' \cdot \mathbf{J}^T(\mathbf{x}') dV',$$

$$(1.5.10)$$

where Eq. $(1.5.4)_2$ has been used twice. ∇' is with respect to $\mathbf{x}'$ and

$$\nabla' \frac{1}{r} = -\nabla \frac{1}{r}. \qquad (1.5.11)$$

The first term on the right-hand side of Eq. (1.5.10) can be converted to a surface integral using the divergence theorem in Appendix C.

When V includes all currents, there are no currents flowing through the surface of V. Therefore, this term vanishes. The second term on the right-hand side of Eq. (1.5.10) also vanishes in magnetostatics because the divergence of a steady-state current distribution is zero. Therefore

$$\nabla \cdot \mathbf{A} = 0. \tag{1.5.12}$$

We also have

$$
\begin{aligned}
\nabla^2 \mathbf{A} &= \frac{\mu_0}{4\pi} \nabla^2 \int_V \frac{\mathbf{J}^T(\mathbf{x}')}{r} dV' \\
&= \frac{\mu_0}{4\pi} \int_V \mathbf{J}^T(\mathbf{x}') \nabla^2 \left(\frac{1}{r} \right) dV' \\
&= \frac{\mu_0}{4\pi} \int_V \mathbf{J}^T(\mathbf{x}') \left[-4\pi \delta(\mathbf{r}) \right] dV' = -\mu_0 \mathbf{J}^T(\mathbf{x}), \quad (1.5.13)
\end{aligned}
$$

where Eq. (1.1.6) has been used. From Eqs. (1.5.9), (1.5.12) and (1.5.13), we have

$$\nabla \times \mathbf{B} = \mu_0 \mathbf{J}^T. \tag{1.5.14}$$

1.6   Current Loops and Magnetic Moments

In a dielectric material without mobile charges, at the microscopic level, molecules carry current loops possessing microscopic magnetic moments. As a simple example, in cylindrical coordinates (r, θ, z) with unit vectors $(\mathbf{e}_r, \mathbf{e}_\theta, \mathbf{e}_z)$, the magnetic moment $\mathbf{m}$ of the circular current loop in Fig. 1.4 is defined by and found to be

$$
\begin{aligned}
\mathbf{m} &= \frac{1}{2} \oint_C \mathbf{r} \times I d\mathbf{l} = \frac{1}{2} \oint_C R \mathbf{e}_r \times I dl \mathbf{e}_\theta \\
&= \frac{1}{2} R I \mathbf{e}_z \oint_C dl = \frac{1}{2} R I \mathbf{e}_z 2\pi R \\
&= I \pi R^2 \mathbf{e}_z = I A \mathbf{e}_z = I \mathbf{A}, \quad (1.6.1)
\end{aligned}
$$

where $A = \pi R^2$ is the area enclosed by the circle and the vector area $\mathbf{A} = A \mathbf{e}_z$.

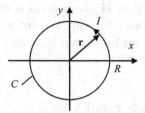

Fig. 1.4. A circular current loop.

Equation (1.6.1) in the form of $\mathbf{m} = I\mathbf{A}$ is also valid for planar current loops of other shapes in general [6]. Consider a planar current loop with an area A and steady current I. In the limit when $A \to 0$, $I \to \infty$ and $IA \to m$, we have

$$\mathbf{m} = \frac{1}{2} \oint_C \mathbf{r} \times I\mathbf{dl}$$

$$= -\frac{1}{2}I \oint_C \mathbf{dl} \times \mathbf{r} = -\frac{1}{2}I \int_A (\mathbf{n} \times \nabla) \times \mathbf{r} dA$$

$$= -\frac{1}{2}I \int_A (-2\mathbf{n}) dA = I \int_A \mathbf{n} dA \to I\mathbf{A}, \qquad (1.6.2)$$

where the following vector identity has been used [7]:

$$\oint_C \mathbf{dl} \times \mathbf{G} = \int_A (\mathbf{n} \times \nabla) \times \mathbf{G} dA. \qquad (1.6.3)$$

A current loop in a magnetic induction field $\mathbf{B}$ experiences a force $\mathbf{f}^M$ which can be calculated from the Biot–Savart law as [6]

$$d\mathbf{f}^M = I\mathbf{dl} \times \mathbf{B},$$

$$\mathbf{f}^M = \oint_C I\mathbf{dl} \times \mathbf{B} = I \int_A (\mathbf{n} \times \nabla) \times \mathbf{B} dA \qquad (1.6.4)$$

$$= I \int_A \mathbf{n} \cdot (\mathbf{B}\nabla) dA$$

$$= I \left(\int_A \mathbf{n} dA \right) \cdot (\mathbf{B}\nabla) \to \mathbf{m} \cdot (\mathbf{B}\nabla), \qquad (1.6.5)$$

or

$$\mathbf{f}^M = \mathbf{m} \cdot (\mathbf{B}\nabla), \quad f_j^M = m_i B_{i,j}. \qquad (1.6.6)$$

The magnetic induction **B** also exerts a moment or couple on a current loop. Taking moment about a point within the area enclosed by the current loop, it can be shown that the moment acting on the current loop is given by [6]

$$\mathbf{c}^M = \oint_C \mathbf{r} \times d\mathbf{f}^M$$

$$= \oint_C \mathbf{r} \times (I d\mathbf{l} \times \mathbf{B}) = \mathbf{m} \times \mathbf{B}. \tag{1.6.7}$$

Equation (1.6.7) shows that $\mathbf{c}^M$ vanishes when **m** is parallel to **B**, and that $\mathbf{c}^M$ tends to align **m** with **B**. The component of **B** along **m** does not contribute to $\mathbf{c}^M$.

For the vector potential **A** and the magnetic induction **B** produced by a current loop with a moment **m** in a vacuum, it can be shown [1,6] that:

$$\mathbf{A} = \frac{\mu_0}{4\pi} \mathbf{m} \times \frac{\mathbf{r}}{r^3},$$

$$\mathbf{B} = -\frac{\mu_0}{4\pi} (\mathbf{m} \cdot \nabla) \frac{\mathbf{r}}{r^3}. \tag{1.6.8}$$

1.7 Magnetization

In a material, the microscopic magnetic moments **m** may be randomly oriented as shown in Fig. 1.5 or aligned to various degrees for different reasons. We define a macroscopic magnetization vector **M** per unit volume by

$$\mathbf{M} = \lim_{\Delta V \to 0} \frac{1}{\Delta V} \sum_{\Delta V} \mathbf{m}. \tag{1.7.1}$$

Consider a finite body occupying a region V whose boundary surface is S with an outward unit normal **n**. It can be shown that, effectively, **M** is equivalent to the following volume magnetization current density $\mathbf{J}^M$ and surface magnetization current density

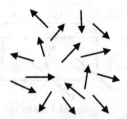

Fig. 1.5.   Microscopic magnetic moments in matter.

$\mathbf{j}^M$ [6]:

$$\mathbf{J}^M = \nabla \times \mathbf{M} = \left(\frac{\partial M_z}{\partial y} - \frac{\partial M_y}{\partial z}\right) \mathbf{e}_1$$

$$+ \left(\frac{\partial M_x}{\partial z} - \frac{\partial M_z}{\partial x}\right) \mathbf{e}_2 + \left(\frac{\partial M_y}{\partial x} - \frac{\partial M_x}{\partial y}\right) \mathbf{e}_3, \quad (1.7.2)$$

$$\mathbf{j}^M = \mathbf{M} \times \mathbf{n}. \tag{1.7.3}$$

For example, part of the z (or $\mathbf{e}_3$) component of Eq. (1.7.2) when the variation of M_x along y is considered can be calculated from Fig. 1.6 as follows. We have, from the figure, for the body current in the z direction near the junction of the two elements:

$$I' \Delta y \Delta z = M_x \Delta x \Delta y \Delta z, \tag{1.7.4}$$

$$I'' \Delta y \Delta z = \left(M_x + \frac{\partial M_x}{\partial y} \Delta y\right) \Delta x \Delta y \Delta z. \tag{1.7.5}$$

Then

$$I' - I'' = -\frac{\partial M_x}{\partial y} \Delta x \Delta y = J_z^M \Delta x \Delta y, \tag{1.7.6}$$

or

$$J_z^M = -\frac{\partial M_x}{\partial y}. \tag{1.7.7}$$

Similarly, when the variation of M_y along x is considered, we have

$$J_z^M = \frac{\partial M_y}{\partial x}. \tag{1.7.8}$$

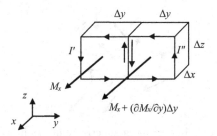

Fig. 1.6.   Contributions to the z component of $\mathbf{J}^M$ from M_x.

Adding Eqs. (1.7.7) and (1.7.8), we obtain the z component of Eq. (1.7.2):

$$J_z^M = \frac{\partial M_y}{\partial x} - \frac{\partial M_x}{\partial y}. \tag{1.7.9}$$

From Fig. 1.6, it can also be seen that if the top surface of the elements is a boundary surface, M_x produces a surface current density

$$j_y^M = -M_x. \tag{1.7.10}$$

Similarly,

$$j_x^M = M_y. \tag{1.7.11}$$

Equations (1.7.10) and (1.7.11) can be written together as

$$\mathbf{j}^M = \mathbf{M} \times \mathbf{n}. \tag{1.7.12}$$

With Eq. (1.7.2), we can write Eq. (1.5.14) as

$$\nabla \times \mathbf{B} = \mu_0(\mathbf{J} + \mathbf{J}^M), \tag{1.7.13}$$

or

$$\nabla \times \left(\frac{1}{\mu_0} \mathbf{B} - \mathbf{M} \right) = \mathbf{J}, \tag{1.7.14}$$

where $\mathbf{J}$ is the current density from origins other than the effective magnetization current. With the introduction of a magnetic field vector $\mathbf{H}$ by

$$\mathbf{H} = \frac{1}{\mu_0}\mathbf{B} - \mathbf{M}, \tag{1.7.15}$$

or

$$\mathbf{B} = \mu_0(\mathbf{H} + \mathbf{M}), \tag{1.7.16}$$

Eq. (1.7.14) can be written as

$$\nabla \times \mathbf{H} = \mathbf{J}. \tag{1.7.17}$$

The integration of the effective magnetization currents in V and on S together is

$$\left(\int_V \nabla \times \mathbf{M}dV + \int_S \mathbf{M} \times \mathbf{n}dS \right)_i$$
$$= \int_V \varepsilon_{ijk}(M_{k,j} + M_{j,k})dV = 0. \tag{1.7.18}$$

The force on the effective magnetization currents in V and on S under $\mathbf{B}$ is

$$\int_V \varepsilon_{ijk} J_j^M B_k dV + \int_S \varepsilon_{ijk} j_j^M B_k dS$$
$$= \int_V M_k B_{k,i} dV = \int_V f_i^M dV, \tag{1.7.19}$$

where, similar to Eq. (1.6.6), the magnetic force on the magnetization $\mathbf{M}$ per unit volume is [6]

$$f_i^M = M_k B_{k,i}, \quad \mathbf{f}^M = \mathbf{M} \cdot (\mathbf{B}\nabla). \tag{1.7.20}$$

It can be verified that the magnetic body force $\mathbf{f}^M + \mathbf{J} \times \mathbf{B} = \mathbf{F}^M$ can be written as the divergence of a magnetostatic stress tensor $\mathbf{T}^M$ as [6]

$$F_l^M = f_l^M + (\mathbf{J} \times \mathbf{B})_l = T_{ml,m}^M, \tag{1.7.21}$$

where

$$T_{ml}^M = B_m H_l + \frac{\mu_0}{2}(M_k M_k - H_k H_k)\delta_{ml}. \tag{1.7.22}$$

1.8 Linear Magnetic Materials

For a linear material,

$$M_k = \chi_{kl}^M H_l, \tag{1.8.1}$$

where χ_{kl}^M is the magnetic susceptibility. Then

$$
\begin{aligned}
B_k &= \mu_0(H_k + M_k) \\
&= \mu_0(\delta_{kl} H_l + \chi_{kl}^M H_l) = \mu_{kl} H_l,
\end{aligned} \tag{1.8.2}
$$

or

$$
\begin{bmatrix} B_1 \\ B_2 \\ B_3 \end{bmatrix} = \begin{bmatrix} \mu_{11} & \mu_{12} & \mu_{13} \\ \mu_{21} & \mu_{22} & \mu_{23} \\ \mu_{31} & \mu_{32} & \mu_{33} \end{bmatrix} \begin{bmatrix} H_1 \\ H_2 \\ H_3 \end{bmatrix}, \tag{1.8.3}
$$

where

$$\mu_{kl} = \mu_0(\delta_{kl} + \chi_{kl}^M) \tag{1.8.4}$$

is the magnetic permeability. For a linear magnetic material, the energy density per unit volume is (see Eq. (1.10.11))

$$\hat{U}(\mathbf{B}) = \frac{1}{2} H_i B_i. \tag{1.8.5}$$

With the following Legendre transform, we introduce a magnetic enthalpy function H^M per unit volume by

$$
\begin{aligned}
H^M(\mathbf{H}) &= \hat{U}(\mathbf{B}) - H_i B_i \\
&= -\frac{1}{2} H_i B_i = -\frac{1}{2} \mu_{ij} H_i H_j.
\end{aligned} \tag{1.8.6}
$$

Then

$$B_i = -\frac{\partial H^M}{\partial H_i} = \mu_{ik} H_k. \tag{1.8.7}$$

In a region where $\mathbf{J} = 0$, Eq. (1.7.17) reduces to

$$\nabla \times \mathbf{H} = 0. \tag{1.8.8}$$

Then a scalar potential ψ can be introduced such that

$$\mathbf{H} = -\nabla\psi, \quad H_i = -\psi_{,i}. \tag{1.8.9}$$

With ψ, Eq. (1.8.7) becomes

$$B_k = \mu_{kl}H_l = -\mu_{kl}\psi_{,l}, \tag{1.8.10}$$

which leads to the following equation for ψ:

$$\nabla \cdot \mathbf{B} = B_{k,k} = (\mu_{kl}H_l)_{,k} = -(\mu_{kl}\psi_{,l})_{,k} = 0. \tag{1.8.11}$$

1.9 Maxwell's Equations

For time-dependent problems, electric and magnetic fields are coupled dynamically and are governed by Maxwell's equations:

$$\nabla \cdot \mathbf{D} = \rho^E,$$

$$\nabla \cdot \mathbf{B} = 0,$$

$$\nabla \times \mathbf{E} = -\frac{\partial \mathbf{B}}{\partial t},$$

$$\nabla \times \mathbf{H} = \mathbf{J} + \frac{\partial \mathbf{D}}{\partial t}. \tag{1.9.1}$$

We also have

$$\mathbf{D} = \varepsilon_0\mathbf{E} + \mathbf{P}, \quad \mathbf{H} = \frac{1}{\mu_0}\mathbf{B} - \mathbf{M}. \tag{1.9.2}$$

In addition, constitutive relations describing material behaviors are needed, e.g.,

$$\mathbf{P} = \mathbf{P}(\mathbf{E}, \mathbf{B}), \quad \mathbf{M} = \mathbf{M}(\mathbf{E}, \mathbf{B}). \tag{1.9.3}$$

The electromagnetic force (Lorentz force) on ρ^E and $\mathbf{J}$ is given by

$$\mathbf{f} = \rho^E\mathbf{E} + \mathbf{J} \times \mathbf{B}. \tag{1.9.4}$$

Equation $(1.9.1)_4$ can be written as

$$\nabla \times \mathbf{H} = \mathbf{J} + \varepsilon_0\frac{\partial \mathbf{E}}{\partial t} + \frac{\partial \mathbf{P}}{\partial t}. \tag{1.9.5}$$

We define a polarization current density by

$$\mathbf{J}^P = \frac{\partial \mathbf{P}}{\partial t} \tag{1.9.6}$$

which satisfies

$$\nabla \cdot \mathbf{J}^P + \frac{\partial \rho^P}{\partial t} = 0. \tag{1.9.7}$$

Taking the time derivative of the first equation and the divergence of the fourth equation in Eq. (1.9.1), respectively, we obtain

$$\frac{\partial}{\partial t}(\nabla \cdot \mathbf{D}) = \frac{\partial \rho^E}{\partial t},$$

$$0 = \nabla \cdot \mathbf{J} + \frac{\partial}{\partial t}(\nabla \cdot \mathbf{D}). \tag{1.9.8}$$

Eliminating the time derivative of the divergence of $\mathbf{D}$, we obtain the conservation of charge as

$$\frac{\partial \rho^E}{\partial t} = -\nabla \cdot \mathbf{J}. \tag{1.9.9}$$

With the introduction of a scalar potential φ and a vector potential $\mathbf{A}$ through

$$\mathbf{B} = \nabla \times \mathbf{A},$$

$$\mathbf{E} = -\nabla\varphi - \frac{\partial \mathbf{A}}{\partial t}, \tag{1.9.10}$$

the following two of Maxwell's equations are satisfied:

$$\nabla \cdot \mathbf{B} = 0,$$

$$\nabla \times \mathbf{E} = -\frac{\partial \mathbf{B}}{\partial t}. \tag{1.9.11}$$

The two remaining ones are

$$\nabla \cdot \mathbf{D} = \rho^E,$$

$$\nabla \times \mathbf{H} = \mathbf{J} + \frac{\partial \mathbf{D}}{\partial t}, \tag{1.9.12}$$

which can be written as equations for φ and $\mathbf{A}$.

As an example of the applications of Maxwell's equations, consider electromagnetic waves in a vacuum. Equation (1.9.1) reduces to

$$\nabla \times \mathbf{E} = -\frac{\partial \mathbf{B}}{\partial t},$$

$$\nabla \times \mathbf{H} = \frac{\partial \mathbf{D}}{\partial t},$$

$$\nabla \cdot \mathbf{D} = 0,$$

$$\nabla \cdot \mathbf{B} = 0. \tag{1.9.13}$$

In a vacuum,

$$\mathbf{D} = \varepsilon_0 \mathbf{E}, \quad \mathbf{B} = \mu_0 \mathbf{H}. \tag{1.9.14}$$

Taking the curl of Eq. $(1.9.13)_1$, we have

$$\nabla \times (\nabla \times \mathbf{E}) = -\nabla \times \frac{\partial \mathbf{B}}{\partial t}. \tag{1.9.15}$$

With the use of the following vector identity:

$$\nabla \times (\nabla \times \mathbf{E}) = \nabla(\nabla \cdot \mathbf{E}) - \nabla^2 \mathbf{E}, \tag{1.9.16}$$

Eq. (1.9.15) becomes

$$\nabla(\nabla \cdot \mathbf{E}) - \nabla^2 \mathbf{E} = -\frac{\partial}{\partial t}(\nabla \times \mathbf{B}), \tag{1.9.17}$$

or

$$\frac{1}{\varepsilon_0}\nabla(\nabla \cdot \mathbf{D}) - \nabla^2 \mathbf{E} = -\frac{\partial}{\partial t}(\nabla \times \mu_0 \mathbf{H}), \tag{1.9.18}$$

where Eq. (1.9.14) has been used. Using Eqs. $(1.9.13)_{2,3}$, we obtain, from Eq. (1.9.18),

$$0 - \nabla^2 \mathbf{E} = -\mu_0 \frac{\partial}{\partial t}\frac{\partial \mathbf{D}}{\partial t}, \tag{1.9.19}$$

or

$$\nabla^2 \mathbf{E} = \varepsilon_0 \mu_0 \frac{\partial^2 \mathbf{E}}{\partial t^2}. \tag{1.9.20}$$

Equation (1.9.20) can be written as

$$\nabla^2 \mathbf{E} = \frac{1}{c^2} \frac{\partial^2 \mathbf{E}}{\partial t^2}, \tag{1.9.21}$$

which is the standard wave equation where

$$c = \frac{1}{\sqrt{\varepsilon_0 \mu_0}} \tag{1.9.22}$$

is the wave speed which in this case is the speed of light in a vacuum. Similarly, it can be shown that

$$\nabla^2 \mathbf{B} = \frac{1}{c^2} \frac{\partial^2 \mathbf{B}}{\partial t^2}. \tag{1.9.23}$$

1.10   Energy and Momentum

Taking the scalar products of the third and fourth equations of Eq. (1.9.1) with $\mathbf{H}$ and $\mathbf{E}$, respectively, we have

$$\mathbf{H} \cdot (\nabla \times \mathbf{E}) = -\mathbf{H} \cdot \frac{\partial \mathbf{B}}{\partial t},$$

$$\mathbf{E} \cdot (\nabla \times \mathbf{H}) = \mathbf{E} \cdot \mathbf{J} + \mathbf{E} \cdot \frac{\partial \mathbf{D}}{\partial t}. \tag{1.10.1}$$

Subtracting the two equations in Eq. (1.10.1) from each other and using the vector identity

$$\nabla \cdot (\mathbf{E} \times \mathbf{H}) = \mathbf{H} \cdot (\nabla \times \mathbf{E}) - \mathbf{E} \cdot (\nabla \times \mathbf{H}), \tag{1.10.2}$$

we obtain Poynting's theorem as

$$\mathbf{E} \cdot \frac{\partial \mathbf{D}}{\partial t} + \mathbf{H} \cdot \frac{\partial \mathbf{B}}{\partial t} + \mathbf{E} \cdot \mathbf{J} = -\nabla \cdot (\mathbf{E} \times \mathbf{H}), \tag{1.10.3}$$

or

$$\frac{\partial \hat{U}}{\partial t} + \mathbf{E} \cdot \mathbf{J} = -\nabla \cdot \mathbf{S}, \tag{1.10.4}$$

where

$$\frac{\partial \hat{U}}{\partial t} = \mathbf{E} \cdot \frac{\partial \mathbf{D}}{\partial t} + \mathbf{H} \cdot \frac{\partial \mathbf{B}}{\partial t}, \tag{1.10.5}$$

$$\mathbf{S} = \mathbf{E} \times \mathbf{H},$$

which suggests that

$$\hat{U} = \hat{U}(\mathbf{D}, \mathbf{B}),$$

$$\mathbf{E} = \frac{\partial \hat{U}}{\partial \mathbf{D}}, \quad \mathbf{H} = \frac{\partial \hat{U}}{\partial \mathbf{B}}. \tag{1.10.6}$$

$\mathbf{S}$ is the Poynting vector or the electromagnetic energy flux per unit area per unit time. With

$$\mathbf{D} = \varepsilon_0 \mathbf{E} + \mathbf{P}, \quad \mathbf{H} = \frac{\mathbf{B}}{\mu_0} - \mathbf{M}, \tag{1.10.7}$$

we have

$$\begin{aligned}
\frac{\partial \hat{U}}{\partial t} &= \mathbf{E} \cdot \left(\varepsilon_0 \frac{\partial \mathbf{E}}{\partial t} + \frac{\partial \mathbf{P}}{\partial t} \right) + \left(\frac{\mathbf{B}}{\mu_0} - \mathbf{M} \right) \cdot \frac{\partial \mathbf{B}}{\partial t} \\
&= \mathbf{E} \cdot \varepsilon_0 \frac{\partial \mathbf{E}}{\partial t} + \frac{\mathbf{B}}{\mu_0} \cdot \frac{\partial \mathbf{B}}{\partial t} + \mathbf{E} \cdot \frac{\partial \mathbf{P}}{\partial t} - \mathbf{M} \cdot \frac{\partial \mathbf{B}}{\partial t} \\
&= \frac{\partial}{\partial t} \left(\frac{\varepsilon_0}{2} \mathbf{E} \cdot \mathbf{E} + \frac{1}{2\mu_0} \mathbf{B} \cdot \mathbf{B} \right) + \mathbf{E} \cdot \frac{\partial \mathbf{P}}{\partial t} - \mathbf{M} \cdot \frac{\partial \mathbf{B}}{\partial t},
\end{aligned} \tag{1.10.8}$$

or

$$\frac{\partial \hat{U}}{\partial t} = \frac{\partial U^F}{\partial t} + \frac{\partial U}{\partial t}, \tag{1.10.9}$$

where

$$U^F = \frac{\varepsilon_0}{2} \mathbf{E} \cdot \mathbf{E} + \frac{1}{2\mu_0} \mathbf{B} \cdot \mathbf{B},$$

$$\frac{\partial U}{\partial t} = \mathbf{E} \cdot \frac{\partial \mathbf{P}}{\partial t} - \mathbf{M} \cdot \frac{\partial \mathbf{B}}{\partial t}. \tag{1.10.10}$$

$\hat{U}$ includes the field energy density U^F and the internal energy density U due to polarization and magnetization. They are all per unit volume. For a linear material,

$$\hat{U} = \frac{1}{2}\mathbf{E} \cdot \mathbf{D} + \frac{1}{2}\mathbf{H} \cdot \mathbf{B}. \qquad (1.10.11)$$

The electromagnetic power on a current loop is given by [6]

$$w^M = \oint_C I\mathbf{dl} \cdot \mathbf{E} = \int_A I\mathbf{n} \cdot (\nabla \times \mathbf{E}) dA$$

$$= -\int_A I\mathbf{n} \cdot \frac{\partial \mathbf{B}}{\partial t} dA = -\left(\int_A I\mathbf{n} dA \right) \cdot \frac{\partial \mathbf{B}}{\partial t}$$

$$\rightarrow -\mathbf{m} \cdot \frac{\partial \mathbf{B}}{\partial t}. \qquad (1.10.12)$$

The electromagnetic body force on ρ^E, $\mathbf{J}$, polarization current, polarization and magnetization is

$$\mathbf{F}^{EM} = \rho^E \mathbf{E} + \mathbf{J} \times \mathbf{B} + \frac{\partial \mathbf{P}}{\partial t} \times \mathbf{B}$$

$$+ \mathbf{P} \cdot \nabla \mathbf{E} + \mathbf{M} \cdot (\mathbf{B}\nabla). \qquad (1.10.13)$$

It can be shown that [6]

$$F_i^{EM} = T_{ki,k}^{EM} - \frac{\partial G_i}{\partial t}, \qquad (1.10.14)$$

where

$$\mathbf{G} = \varepsilon_0 \mathbf{E} \times \mathbf{B}, \quad G_i = \varepsilon_0 \varepsilon_{ijk} E_j B_k \qquad (1.10.15)$$

is the momentum density vector and

$$T_{ki}^{EM} = D_k E_i + B_k H_i$$

$$- \frac{1}{2}(\varepsilon_0 E_j E_j + \mu_0 H_j H_j - \mu_0 M_j M_j)\delta_{ki}. \qquad (1.10.16)$$

1.11 Boundary Conditions

For boundary conditions we need the following integral forms of Maxwell's equations:

$$\int_S \mathbf{n}\cdot\mathbf{D}dS = \int_V \rho^E dV, \qquad (1.11.1)$$

$$\int_S \mathbf{n}\cdot\mathbf{B}dS = 0, \qquad (1.11.2)$$

$$\oint_C \mathbf{H}\cdot d\mathbf{x} = \frac{\partial}{\partial t}\int_S \mathbf{n}\cdot\mathbf{D}dS + \int_S \mathbf{n}\cdot\mathbf{J}dS, \qquad (1.11.3)$$

$$\oint_C \mathbf{E}\cdot d\mathbf{x} = -\frac{\partial}{\partial t}\int_S \mathbf{n}\cdot\mathbf{B}dS. \qquad (1.11.4)$$

We consider continuity or jump conditions in general at a fixed interface between two stationary materials. $\partial\mathbf{D}/\partial t$ and $\partial\mathbf{B}/\partial t$ are assumed to be bounded near the interface. The interface may carry a surface charge density σ^E per unit area and a surface current density $\mathbf{j}$ per unit length and unit time.

For Eq. (1.11.1), we construct a pillbox at an interface as shown in Fig. 1.7 [9]. $\mathbf{n}$ in the unit normal of the interface pointing from material "$-$" to material "$+$". The top and bottom surface areas of the pillbox is S. The thickness δ of the pillbox is infinitesimal. So are the lateral cylindrical surface area and the volume of the pillbox which are proportional to δ. The application of Eq. (1.11.1) to the pillbox yields

$$\mathbf{n}\cdot\mathbf{D}^+ S + (-\mathbf{n})\cdot\mathbf{D}^- S \cong \sigma^E S, \qquad (1.11.5)$$

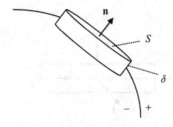

Fig. 1.7. A pillbox at an interface between two materials.

where the contributions from the infinitesimal lateral surface and the infinitesimal volume of the pillbox have been neglected. Equation (1.11.5) can be further written as

$$\mathbf{n} \cdot (\mathbf{D}^+ - \mathbf{D}^-) = \sigma^E, \tag{1.11.6}$$

or

$$\mathbf{n} \cdot [\mathbf{D}] = \sigma^E, \tag{1.11.7}$$

where the jump of a field ϕ across the interface is denoted by square brackets:

$$[\phi] = \phi^+ - \phi^-. \tag{1.11.8}$$

Similarly, the continuity condition corresponding to Eq. (1.11.2) is

$$\mathbf{n} \cdot [\mathbf{B}] = 0. \tag{1.11.9}$$

For Eq. (1.11.3), consider the stationary closed curved (contour) at the interface in Fig. 1.8 [9]. $\mathbf{t}_1$ and $\mathbf{t}_2$ are orthogonal unit tangent vectors at the interface. $\delta \ll L$ and is infinitesimal. So is the area S within the closed curve. $\mathbf{t}_1$ is the unit normal of S. We Apply Eq. (1.11.3) to the closed curve in Fig. 1.8. The integration of $\partial \mathbf{D}/\partial t$ over S vanishes because S is infinitesimal and $\partial \mathbf{D}/\partial t$ is bounded. Hence

$$-H_2^+ L + H_2^- L \cong j_1 L, \tag{1.11.10}$$

or

$$-H_2^+ + H_2^- \cong j_1. \tag{1.11.11}$$

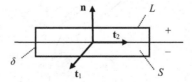

Fig. 1.8.   A closed curve at an interface between two materials.

If a closed curve in the plane containing $\mathbf{n}$ and $\mathbf{t}_1$ is used, we have

$$H_1^+ - H_1^- \cong j_2. \tag{1.11.12}$$

Equations (1.11.11) and (1.11.12) can be written together as

$$\mathbf{n} \times [\mathbf{H}] = \mathbf{j}. \tag{1.11.13}$$

Similarly, Eq. (1.11.4) leads to

$$\mathbf{n} \times [\mathbf{E}] = 0. \tag{1.11.14}$$

1.12 Variational Formulation

Consider the following Lagrangian density L [8]

$$L(\varphi, \mathbf{A}, \mathbf{P}, \mathbf{M}) = \frac{1}{2}\left(\varepsilon_0 \mathbf{E} \cdot \mathbf{E} - \frac{1}{\mu_0}\mathbf{B} \cdot \mathbf{B}\right)$$
$$+ \mathbf{E} \cdot \mathbf{P} + \mathbf{B} \cdot \mathbf{M} - W(\mathbf{P}, \mathbf{M}), \tag{1.12.1}$$

where

$$\mathbf{B} = \nabla \times \mathbf{A},$$
$$\mathbf{E} = -\nabla\varphi - \frac{\partial \mathbf{A}}{\partial t}, \tag{1.12.2}$$

with which two of the four Maxwell's equations are satisfied. We have

$$\delta L = D_{i,i}\delta\varphi + \left(\frac{\partial D_k}{\partial t} - \varepsilon_{kji}H_{i,j}\right)\partial A_k$$
$$- (D_i\delta\varphi)_{,i} - \frac{\partial}{\partial t}(D_i\delta A_i) - (H_i\varepsilon_{ijk}\delta A_k)_{,j}, \tag{1.12.3}$$

where we have denoted

$$\mathbf{D} = \varepsilon_0 \mathbf{E} + \mathbf{P}, \quad \mathbf{H} = \frac{\mathbf{B}}{\mu_0} - \mathbf{M}, \tag{1.12.4}$$

$$\mathbf{E} = \frac{\partial W}{\partial \mathbf{P}}, \quad \mathbf{B} = \frac{\partial W}{\partial \mathbf{M}}. \tag{1.12.5}$$

Consider the following functional Π based on L:

$$\Pi = \int_{t_0}^{t_1} dt \int_V L dV. \tag{1.12.6}$$

Clearly, the stationary condition of Π implies the two remaining ones of Maxwell's equations:

$$D_{i,i} = 0, \quad \varepsilon_{kji} H_{i,j} = \frac{\partial D_k}{\partial t}. \tag{1.12.7}$$

We are more interested in a variational principle with $\mathbf{E}$ and $\mathbf{B}$ as independent fields. For this purpose we introduce V through the following Legendre transform of W:

$$V(\mathbf{E}, \mathbf{B}) = W(\mathbf{P}, \mathbf{M}) - \mathbf{E} \cdot \mathbf{P} - \mathbf{B} \cdot \mathbf{M}. \tag{1.12.8}$$

Then

$$\mathbf{P} = -\frac{\partial V}{\partial \mathbf{E}}, \quad \mathbf{M} = -\frac{\partial V}{\partial \mathbf{B}}. \tag{1.12.9}$$

Consider the following Lagrangian density:

$$L(\varphi, \mathbf{A}) = \frac{1}{2} \left(\varepsilon_0 \mathbf{E} \cdot \mathbf{E} - \frac{1}{\mu_0} \mathbf{B} \cdot \mathbf{B} \right) - V(\mathbf{E}, \mathbf{B}). \tag{1.12.10}$$

Then

$$\delta L = D_{i,i} \delta\varphi + \left(\frac{\partial D_k}{\partial t} - \varepsilon_{kji} H_{i,j} \right) \partial A_k$$

$$- (D_i \delta\varphi)_{,i} - \frac{\partial}{\partial t} (D_i \delta A_i) - (H_i \varepsilon_{ijk} \delta A_k)_{,j}. \tag{1.12.11}$$

Assuming that $\delta\mathbf{A}$ vanishes at t_0 and t_1, we have

$$\delta\Pi = \int_{t_0}^{t_1} dt \int_V \left[D_{i,i} \delta\varphi + \left(\frac{\partial D_k}{\partial t} - \varepsilon_{kji} H_{i,j} \right) \partial A_k \right] dV$$

$$- \int_{t_0}^{t_1} dt \int_S (n_i D_i \delta\varphi + \varepsilon_{kij} H_i n_j \delta A_k) dS. \tag{1.12.12}$$

Hence the stationary condition of Π is

$$D_{i,i} = 0,$$

$$\varepsilon_{kji} H_{i,j} = \frac{\partial D_k}{\partial t}. \tag{1.12.13}$$

Equation (1.12.12) also suggests the following possible boundary conditions on S:

$$\varphi \quad \text{or} \quad \mathbf{n} \cdot \mathbf{D},$$

$$\mathbf{A} \quad \text{or} \quad \mathbf{n} \times \mathbf{H}. \tag{1.12.14}$$

1.13 Lorentz Transformation

Consider two inertial reference frames $(\mathbf{x}, t)$ and $(\mathbf{x}', t')$ or R_G and R_C as shown in Fig. 1.9. R_G is fixed. R_C is moving with a constant velocity v in the x_1 direction relative to R_G (standard configuration).

The Lorentz transformation between the two reference frames is

$$x_1' = \frac{x_1 - vt}{\sqrt{1 - v^2/c^2}} = \alpha(x_1 - vt),$$

$$x_2' = x_2, \quad x_3' = x_3, \tag{1.13.1}$$

$$t' = \frac{t - vx_1/c^2}{\sqrt{1 - v^2/c^2}} = \alpha \left(t - \frac{vx_1}{c^2} \right),$$

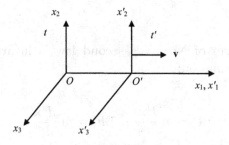

Fig. 1.9.   Two inertial frames.

where

$$\alpha = \frac{1}{\sqrt{1 - v^2/c^2}}. \tag{1.13.2}$$

The inverse transformation is given by

$$x_1 = \alpha(x_1' + vt'),$$
$$x_2 = x_2', \quad x_3 = x_3',$$
$$t = \alpha\left(t' + \frac{vx_1'}{c^2}\right). \tag{1.13.3}$$

The Lorentz transformation preserves

$$x_1^2 - c^2t^2 = (x_1')^2 - c^2(t')^2. \tag{1.13.4}$$

When $v \ll c$, Eq. (13.1) reduces to the Galilean transformation below:

$$x_1' = x_1 - vt,$$
$$x_2' = x_2, \quad x_3' = x_3,$$
$$t' = t. \tag{1.13.5}$$

The classical form of Newton's second law is invariant under the Galilean transformation:

$$\Sigma\mathbf{F} = m\frac{d^2\mathbf{x}}{dt^2}, \quad \Sigma\mathbf{F}' = m'\frac{d^2\mathbf{x}'}{dt'^2},$$
$$\mathbf{F}' = \mathbf{F}, \quad m' = m. \tag{1.13.6}$$

1.14   Fields in a Moving Body

In the fixed inertial reference frame R_G, Maxwell's equations are

$$\varepsilon_{ijk}\frac{\partial E_k}{\partial x_j} = -\frac{\partial B_i}{\partial t},$$

$$\frac{\partial B_k}{\partial x_k} = 0, \tag{1.14.1}$$

$$\varepsilon_{ijk}\frac{\partial H_k}{\partial x_j} = \frac{\partial D_i}{\partial t} + J_i,$$

$$\frac{\partial D_k}{\partial x_k} = \rho^E. \tag{1.14.2}$$

Consider Eq. (1.14.1) first. Under the Lorentz transformation in Eq. (1.13.1), we can write Eq. (1.14.1) as [6]

$$\varepsilon_{ijk}\left(\frac{\partial E_k}{\partial x'_l}\frac{\partial x'_l}{\partial x_j} + \frac{\partial E_k}{\partial t'}\frac{\partial t'}{\partial x_j}\right) + \left(\frac{\partial B_i}{\partial x'_l}\frac{\partial x'_l}{\partial t} + \frac{\partial B_i}{\partial t'}\frac{\partial t'}{\partial t}\right) = 0,$$

$$\tag{1.14.3}$$

$$\frac{\partial B_k}{\partial x'_l}\frac{\partial x'_l}{\partial x_k} + \frac{\partial B_k}{\partial t'}\frac{\partial t'}{\partial x_k} = 0. \tag{1.14.4}$$

From Eq. (1.13.1) we obtain

$$\frac{\partial x'_1}{\partial x_1} = \alpha, \quad \frac{\partial x'_1}{\partial t} = -\alpha v, \quad \frac{\partial x'_2}{\partial x_2} = 1, \quad \frac{\partial x'_3}{\partial x_3} = 1,$$

$$\frac{\partial t'}{\partial t} = \alpha, \quad \frac{\partial t'}{\partial x_1} = -\alpha\frac{v}{c^2}, \tag{1.14.5}$$

and the other ten derivatives vanish. Now, writing Eq. (1.14.1) out
fully, we have

$$\frac{\partial B_1}{\partial x_1} + \frac{\partial B_2}{\partial x_2} + \frac{\partial B_3}{\partial x_3} = 0,$$

$$\frac{\partial E_3}{\partial x_2} - \frac{\partial E_2}{\partial x_3} + \frac{\partial B_1}{\partial t} = 0, \tag{1.14.6}$$

$$\frac{\partial E_1}{\partial x_3} - \frac{\partial E_3}{\partial x_1} + \frac{\partial B_2}{\partial t} = 0,$$

$$\frac{\partial E_2}{\partial x_1} - \frac{\partial E_1}{\partial x_2} + \frac{\partial B_3}{\partial t} = 0. \tag{1.14.7}$$

Rewriting Eqs. (1.14.6) and (1.14.7) using Eq. (1.14.5), we obtain

$$\alpha\frac{\partial B_1}{\partial x_1'} - \alpha\frac{v}{c^2}\frac{\partial B_1}{\partial t'} + \frac{\partial B_2}{\partial x_2'} + \frac{\partial B_3}{\partial x_3'} = 0, \tag{1.14.8}$$

$$\frac{\partial E_3}{\partial x_2'} - \frac{\partial E_2}{\partial x_3'} + \alpha\frac{\partial B_1}{\partial t'} - \alpha v\frac{\partial B_1}{\partial x_1'} = 0, \tag{1.14.9}$$

$$\frac{\partial E_1}{\partial x_3'} - \frac{\partial}{\partial x_1'}(\alpha E_3 + \alpha v B_2) + \frac{\partial}{\partial t'}\left(\alpha B_2 + \alpha\frac{v}{c^2}E_3\right) = 0,$$

$$\tag{1.14.10}$$

$$\frac{\partial}{\partial x_1'}(\alpha E_2 - \alpha v B_3) - \frac{\partial E_1}{\partial x_2'} + \frac{\partial}{\partial t'}\left(\alpha B_3 - \alpha\frac{v}{c^2}E_2\right) = 0.$$

$$\tag{1.14.11}$$

Multiplying Eq. (1.14.8) by α, Eq. (1.14.9) by $\alpha v/c^2$, and adding
them, we arrive at

$$\frac{\partial B_1}{\partial x_1'} + \frac{\partial}{\partial x_2'}\left(\alpha B_2 + \alpha\frac{v}{c^2}E_3\right) + \frac{\partial}{\partial x_3'}\left(\alpha B_3 - \alpha\frac{v}{c^2}E_2\right) = 0. \tag{1.14.12}$$

Multiplying Eq. (1.14.8) by αv, Eq. (1.14.9) by α, and adding them, we have

$$\frac{\partial}{\partial x'_2}(\alpha E_3 + \alpha v B_2) - \frac{\partial}{\partial x'_3}(\alpha E_2 - \alpha v B_3) + \frac{\partial B_1}{\partial t'} = 0. \qquad (1.14.13)$$

From Eqs. (1.14.10)–(1.14.13) we identify

$$E'_1 = E_1, \quad B'_1 = B_1,$$

$$E'_2 = \alpha(E_2 - vB_3), \quad B'_2 = \alpha\left(B_2 + \frac{v}{c^2}E_3\right),$$

$$E'_3 = \alpha(E_3 + vB_2), \quad B'_3 = \alpha\left(B_3 - \frac{v}{c^2}E_2\right). \qquad (1.14.14)$$

Then Eqs. (1.14.10)–(1.14.13) can be written as

$$\frac{\partial B'_1}{\partial x'_1} + \frac{\partial B'_2}{\partial x'_2} + \frac{\partial B'_3}{\partial x'_3} = 0,$$

$$\frac{\partial E'_3}{\partial x'_2} - \frac{\partial E'_2}{\partial x'_3} + \frac{\partial B'_1}{\partial t'} = 0, \qquad (1.14.15)$$

$$\frac{\partial E'_1}{\partial x'_3} - \frac{\partial E'_3}{\partial x'_1} + \frac{\partial B'_2}{\partial t'} = 0,$$

$$\frac{\partial E'_2}{\partial x'_1} - \frac{\partial E'_1}{\partial x'_2} + \frac{\partial B'_3}{\partial t'} = 0. \qquad (1.14.16)$$

Equations (1.14.15) and (1.14.16) have the same form as Eqs. (1.14.6) and (1.14.7). Thus Eq. (1.14.1) is invariant under the Lorentz transformation. Equation (1.14.14) can be written as

$$\mathbf{E}'_p = \mathbf{E}_p, \quad \mathbf{E}'_n = \alpha(\mathbf{E}_n + \mathbf{v} \times \mathbf{B}_n),$$

$$\mathbf{B}'_p = \mathbf{B}_p, \quad \mathbf{B}'_n = \alpha\left(\mathbf{B}_n - \frac{1}{c^2}\mathbf{v} \times \mathbf{E}_n\right), \qquad (1.14.17)$$

where the subscripts p and n, respectively, stand for parallel and normal to $\mathbf{v}$. Since $\mathbf{v} \times \mathbf{E}_p = 0$ and $\mathbf{v} \times \mathbf{B}_p = 0$, Eq. (1.14.17) can also be written as

$$\mathbf{E}'_p = \mathbf{E}_p + \mathbf{v} \times \mathbf{B}_p,$$
$$\mathbf{E}'_n = \alpha(\mathbf{E}_n + \mathbf{v} \times \mathbf{B}_n), \tag{1.14.18}$$

$$\mathbf{B}'_p = \mathbf{B}_p - \frac{1}{c^2} \mathbf{v} \times \mathbf{E}_p,$$
$$\mathbf{B}'_n = \alpha \left(\mathbf{B}_n - \frac{1}{c^2} \mathbf{v} \times \mathbf{E}_n \right). \tag{1.14.19}$$

Let us now confine our attention to Eq. (1.14.2). When written out fully, Eq. (1.14.2) takes the following form:

$$\frac{\partial D_1}{\partial x_1} + \frac{\partial D_2}{\partial x_2} + \frac{\partial D_3}{\partial x_3} = \rho^E,$$

$$\frac{\partial H_3}{\partial x_2} - \frac{\partial H_2}{\partial x_3} - \frac{\partial D_1}{\partial t} = J_1, \tag{1.14.20}$$

$$\frac{\partial H_1}{\partial x_3} - \frac{\partial H_3}{\partial x_1} - \frac{\partial D_2}{\partial t} = J_2,$$

$$\frac{\partial H_2}{\partial x_1} - \frac{\partial H_1}{\partial x_2} - \frac{\partial D_3}{\partial t} = J_3. \tag{1.14.21}$$

In the primed reference frame, we have

$$\alpha \frac{\partial D_1}{\partial x'_1} - \alpha \frac{v}{c^2} \frac{\partial D_1}{\partial t'} + \frac{\partial D_2}{\partial x'_2} + \frac{\partial D_3}{\partial x'_3} = \rho^E, \tag{1.14.22}$$

$$\frac{\partial H_3}{\partial x'_2} - \frac{\partial H_2}{\partial x'_3} - \alpha \frac{\partial D_1}{\partial t'} + \alpha v \frac{\partial D_1}{\partial x'_1} = J_1, \tag{1.14.23}$$

$$\frac{\partial H_1}{\partial x'_3} - \frac{\partial}{\partial x'_1}(\alpha H_3 - \alpha v D_2) - \frac{\partial}{\partial t'} \left(\alpha D_2 - \alpha \frac{v}{c^2} H_3 \right) = J_2,$$
$$\tag{1.14.24}$$

$$\frac{\partial}{\partial x'_1}(\alpha H_2 + \alpha v D_3) - \frac{\partial H_1}{\partial x'_2} - \frac{\partial}{\partial t'} \left(\alpha D_3 + \alpha \frac{v}{c^2} H_2 \right) = J_3.$$
$$\tag{1.14.25}$$

Multiplying Eq. (1.14.22) by α, Eq. (1.14.23) by $\alpha v/c^2$, and subtracting the second from the first, we obtain

$$\frac{\partial D_1}{\partial x_1'} + \frac{\partial}{\partial x_2'}\left(\alpha D_2 - \alpha \frac{v}{c^2} H_3\right)$$

$$+ \frac{\partial}{\partial x_3'}\left(\alpha D_3 + \alpha \frac{v}{c^2} H_2\right) = \alpha \rho^E - \alpha \frac{v}{c^2} J_1. \quad (1.14.26)$$

Multiplying Eq. (1.14.22) by αv, Eq. (1.14.23) by α, and subtracting the first from the second, we have

$$\frac{\partial}{\partial x_2'}(\alpha H_3 - \alpha v D_2) - \frac{\partial}{\partial x_3'}(\alpha H_2 + \alpha v D_3)$$

$$- \frac{\partial D_1}{\partial t'} = \alpha J_1 - \alpha v \rho^E. \quad (1.14.27)$$

We identify

$$D_1' = D_1, \quad H_1' = H_1,$$

$$D_2' = \alpha\left(D_2 - \frac{v}{c^2} H_3\right), \quad H_2' = \alpha(H_2 + v D_3), \quad (1.14.28)$$

$$D_3' = \alpha\left(D_3 + \frac{v}{c^2} H_2\right), \quad H_3' = \alpha(H_3 - v D_2),$$

$$\rho'^E = \alpha\left(\rho^E - \frac{v}{c^2} J_1\right),$$

$$J_1' = \alpha(J_1 - \rho^E v), \quad J_2' = J_2, \quad J_3' = J_3. \quad (1.14.29)$$

Then Eqs. (1.14.24)–(1.14.27) become

$$\frac{\partial D_1'}{\partial x_1'} + \frac{\partial D_2'}{\partial x_2'} + \frac{\partial D_3'}{\partial x_3'} = \rho'^E,$$

$$\frac{\partial H_3'}{\partial x_2'} - \frac{\partial H_2'}{\partial x_3'} - \frac{\partial D_1'}{\partial t'} = J_1', \quad (1.14.30)$$

$$\frac{\partial H'_1}{\partial x'_3} - \frac{\partial H'_3}{\partial x'_1} - \frac{\partial D'_2}{\partial t'} = J'_2,$$

$$\frac{\partial H'_2}{\partial x'_1} - \frac{\partial H'_1}{\partial x'_2} - \frac{\partial D'_3}{\partial t'} = J'_3. \tag{1.14.31}$$

Thus Eq. (1.14.2) is invariant under the Lorentz transformation. Equations (1.14.28) and (1.14.29) can be written as

$$\mathbf{D}'_p = \mathbf{D}_p, \quad \mathbf{D}'_n = \alpha\left(\mathbf{D}_n + \frac{1}{c^2}\mathbf{v}\times\mathbf{H}_n\right),$$

$$\mathbf{H}'_p = \mathbf{H}_p, \quad \mathbf{H}'_n = \alpha(\mathbf{H}_n - \mathbf{v}\times\mathbf{D}_n), \tag{1.14.32}$$

$$\rho'^E = \alpha\left(\rho^E - \frac{1}{c^2}\mathbf{v}\cdot\mathbf{J}\right),$$

$$\mathbf{J}'_p = \alpha(\mathbf{J}_p - \rho^E\mathbf{v}), \quad \mathbf{J}'_n = \mathbf{J}_n. \tag{1.14.33}$$

Equation (1.14.32) can also be written as

$$\mathbf{D}'_p = \mathbf{D}_p + \frac{1}{c^2}\mathbf{v}\times\mathbf{H}_p,$$

$$\mathbf{D}'_n = \alpha\left(\mathbf{D}_n + \frac{1}{c^2}\mathbf{v}\times\mathbf{H}_n\right), \tag{1.14.34}$$

$$\mathbf{H}'_p = \mathbf{H}_p - \mathbf{v}\times\mathbf{D}_p,$$

$$\mathbf{H}'_n = \alpha(\mathbf{H}_n - \mathbf{v}\times\mathbf{D}_n). \tag{1.14.35}$$

From the invariance of the following relations under the Lorentz transformation:

$$\mathbf{D} = \varepsilon_0\mathbf{E} + \mathbf{P}, \quad \mathbf{H} = \frac{\mathbf{B}}{\mu_0} - \mathbf{M}, \tag{1.14.36}$$

we identify

$$\mathbf{P}'_p = \mathbf{P}_p - \frac{1}{c^2}\mathbf{v}\times\mathbf{M}_p,$$

$$\mathbf{P}'_n = \alpha\left(\mathbf{P}_n - \frac{1}{c^2}\mathbf{v}\times\mathbf{M}_n\right), \tag{1.14.37}$$

$$\mathbf{M}'_p = \mathbf{M}_p + \mathbf{v} \times \mathbf{E}_p,$$
$$\mathbf{M}'_n = \alpha(\mathbf{M}_n + \mathbf{v} \times \mathbf{E}_n), \tag{1.14.38}$$

where we have used

$$c^2 = \frac{1}{\varepsilon_0 \mu_0}. \tag{1.14.39}$$

If $v \ll c$ and $\alpha \cong 1$, the fields in R_C are related to those in R_G approximately by

$$\mathbf{E}' \cong \mathbf{E} + \mathbf{v} \times \mathbf{B},$$

$$\mathbf{B}' \cong \mathbf{B} - \frac{1}{c^2}\mathbf{v} \times \mathbf{E}, \tag{1.14.40}$$

$$\mathbf{D}' \cong \mathbf{D} + \frac{1}{c^2}\mathbf{v} \times \mathbf{H},$$

$$\mathbf{H}' \cong \mathbf{H} - \mathbf{v} \times \mathbf{D}, \tag{1.14.41}$$

$$\rho'^E \cong \rho^E - \frac{1}{c^2}\mathbf{v} \cdot \mathbf{J},$$

$$\mathbf{J}' \cong \mathbf{J} - \rho^E \mathbf{v}, \tag{1.14.42}$$

$$\mathbf{P}' \cong \mathbf{P} - \frac{1}{c^2}\mathbf{v} \times \mathbf{M},$$

$$\mathbf{M}' \cong \mathbf{M} + \mathbf{v} \times \mathbf{P}. \tag{1.14.43}$$

Further approximations may be made in Eqs. (1.14.40)–(1.14.43) by dropping the terms multiplied with $1/c^2$:

$$\mathbf{E}' \cong \mathbf{E} + \mathbf{v} \times \mathbf{B},$$
$$\mathbf{B}' \cong \mathbf{B}, \tag{1.14.44}$$

$$\mathbf{D}' \cong \mathbf{D},$$
$$\mathbf{H}' \cong \mathbf{H} - \mathbf{v} \times \mathbf{D}, \tag{1.14.45}$$

$$\rho'^E \cong \rho^E,$$
$$\mathbf{J}' \cong \mathbf{J} - \rho^E \mathbf{v}, \tag{1.14.46}$$

$$\mathbf{P}' \cong \mathbf{P},$$
$$\mathbf{M}' \cong \mathbf{M} + \mathbf{v} \times \mathbf{P}. \tag{1.14.47}$$

Chapter 2

Continuum Kinematics

In this chapter, we develop the nonlinear kinematics of a continuum with large deformations. The two-point Cartesian tensor notation is used [10]. The chapter is not meant to be a complete treatment. Only the results needed for the rest of the book are presented.

2.1 Coordinate Systems

Consider a deformable continuum which, in the reference configuration at time t_0, occupies a region V with a boundary surface S (see Fig. 2.1). $\mathbf{N}$ is the outward unit normal of S. The position of a material point in this state is denoted by a position vector $\mathbf{X} = X_K \mathbf{I}_K$ in a rectangular coordinate system. X_K denote the reference or material coordinates of the material point. They are a continuous labeling of material particles so that they are identifiable. At time t, the body occupies a region v with a boundary surface s and an outward unit normal $\mathbf{n}$. The current position of the material point associated with $\mathbf{X}$ is given by $\mathbf{y} = y_k \mathbf{i}_k$ which denotes the present or spatial coordinates of the material point. $\mathbf{u}$ is the displacement vector.

The coordinate systems are assumed to be orthogonal, i.e.,

$$\mathbf{I}_k \cdot \mathbf{I}_l = \delta_{kl}, \quad \mathbf{I}_K \cdot \mathbf{I}_L = \delta_{KL}, \tag{2.1.1}$$

41

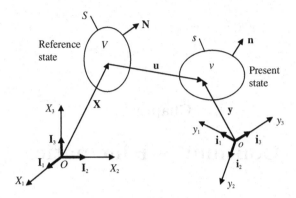

Fig. 2.1.   Motion of a continuum and coordinate systems.

where δ_{kl} and δ_{KL} are the Kronecker delta. In matrix notation,

$$[\delta_{kl}] = [\delta_{KL}] = \begin{bmatrix} 1 & 0 & 0 \\ 0 & 1 & 0 \\ 0 & 0 & 1 \end{bmatrix}. \qquad (2.1.2)$$

The transformation coefficients between the unit vectors of the two coordinate systems are denoted by

$$\mathbf{i}_k \cdot \mathbf{I}_L = \delta_{kL} = \delta_{Lk} = \mathbf{I}_L \cdot \mathbf{i}_k. \qquad (2.1.3)$$

In the rest of this book, the two coordinate systems are chosen to be coincident, i.e.,

$$o = O, \quad \mathbf{i}_1 = \mathbf{I}_1, \quad \mathbf{i}_2 = \mathbf{I}_2, \quad \mathbf{i}_3 = \mathbf{I}_3. \qquad (2.1.4)$$

Then δ_{kL} becomes the Kronecker delta. A vector can be resolved into rectangular components in different coordinate systems. For example, we can also write

$$\mathbf{y} = y_K \mathbf{I}_K, \qquad (2.1.5)$$

with

$$y_M = \delta_{Mi} y_i. \qquad (2.1.6)$$

2.2 Motion of a Continuum

The motion of the body is described by

$$y_i = y_i(\mathbf{X}, t). \tag{2.2.1}$$

The displacement vector $\mathbf{u}$ of a material point is related to $\mathbf{X}$ and $\mathbf{y}$ through

$$\mathbf{y} = \mathbf{X} + \mathbf{u}, \tag{2.2.2}$$

or

$$y_i = \delta_{iM}(X_M + u_M). \tag{2.2.3}$$

An infinitesimal material line element $\mathbf{dX}$ at t_0 deforms into the following line element at t:

$$dy_i|_{t \text{ fixed}} = y_{i,K} dX_K, \tag{2.2.4}$$

where the deformation gradient

$$y_{k,K} = \delta_{kK} + u_{k,K} \tag{2.2.5}$$

is a two-point tensor in the sense that it depends on both of the two coordinate systems for $\mathbf{y}$ and $\mathbf{X}$. The following determinant is called the Jacobian of the deformation:

$$J = \det(y_{k,K}) = \varepsilon_{ijk} y_{i,1} y_{j,2} y_{k,3}$$

$$= \varepsilon_{KLM} y_{1,K} y_{2,L} y_{3,M} = \frac{1}{6} \varepsilon_{klm} \varepsilon_{KLM} y_{k,K} y_{l,L} y_{m,M}, \tag{2.2.6}$$

where ε_{klm} (or ε_{KLM}) is the permutation tensor, and

$$\varepsilon_{ijk} = \mathbf{i}_i \cdot (\mathbf{i}_j \times \mathbf{i}_k)$$

$$= \begin{cases} 1 & i,j,k = 1,2,3; \quad 2,3,1; \quad 3,1,2, \\ -1 & i,j,k = 3,2,1; \quad 2,1,3; \quad 1,3,2, \\ 0 & \text{otherwise.} \end{cases} \tag{2.2.7}$$

The following relationship exists ($\varepsilon - \delta$ identity) [11]:

$$\varepsilon_{ijk}\varepsilon_{pqr} = \begin{vmatrix} \delta_{ip} & \delta_{iq} & \delta_{ir} \\ \delta_{jp} & \delta_{jq} & \delta_{jr} \\ \delta_{kp} & \delta_{kq} & \delta_{kr} \end{vmatrix}. \tag{2.2.8}$$

As a special case, when $i = p$, Eq. (2.2.8) reduces to

$$\varepsilon_{ijk}\varepsilon_{iqr} = \delta_{jq}\delta_{kr} - \delta_{jr}\delta_{kq}. \tag{2.2.9}$$

With Eq. (2.2.8), it can be shown from Eq. (2.2.6) that [12]

$$J = \frac{1}{6}\left[2\frac{\partial y_K}{\partial X_L}\frac{\partial y_L}{\partial X_M}\frac{\partial y_M}{\partial X_K} - 3\frac{\partial y_K}{\partial X_K}\frac{\partial y_L}{\partial X_M}\frac{\partial y_M}{\partial X_L} + \left(\frac{\partial y_M}{\partial X_M}\right)^3\right]. \tag{2.2.10}$$

It can be verified directly that for any L, M and N the following is true:

$$\varepsilon_{ijk}y_{i,L}y_{j,M}y_{k,N} = J\varepsilon_{LMN}. \tag{2.2.11}$$

From Eq. (2.2.11) the following can be shown:

$$\varepsilon_{ijk}y_{j,M}y_{k,N} = J\varepsilon_{LMN}X_{L,i},$$
$$\varepsilon_{ijk}y_{k,N} = J\varepsilon_{LMN}X_{L,i}X_{M,j}. \tag{2.2.12}$$

Proof. Multiplying both sides of Eq. (2.2.11) by $X_{L,r}$, we have

$$\varepsilon_{ijk}y_{i,L}X_{L,r}y_{j,M}y_{k,N} = J\varepsilon_{LMN}X_{L,r}. \tag{2.2.13}$$

Then

$$\varepsilon_{ijk}\delta_{ir}y_{j,M}y_{k,N} = J\varepsilon_{LMN}X_{L,r}, \tag{2.2.14}$$
$$\varepsilon_{rjk}y_{j,M}y_{k,N} = J\varepsilon_{LMN}X_{L,r}. \tag{2.2.15}$$

Replacing the index r by i gives Eq. $(2.2.12)_1$. Similarly, Eq. $(2.2.12)_2$ can be proved by multiplying both sides of Eq. $(2.2.12)_1$ by $X_{M,p}$.

Then the following relationship can be established:

$$\varepsilon_{ijk}\varepsilon_{LMN}y_{j,M}y_{k,N} = 2JX_{L,i}. \tag{2.2.16}$$

Proof. Multiplying both sides of Eq. $(2.2.12)_1$ by ε_{PMN} gives

$$\varepsilon_{ijk}\varepsilon_{PMN}y_{j,M}y_{k,N} = J\varepsilon_{LMN}\varepsilon_{PMN}X_{L,i} = J\varepsilon_{MNL}\varepsilon_{MNP}X_{L,i}$$
$$= J(\delta_{NN}\delta_{LP} - \delta_{NP}\delta_{LN})X_{L,i}$$
$$= J(3\delta_{LP} - \delta_{LP})X_{L,i} = 2JX_{P,i}, \tag{2.2.17}$$

where Eq. (2.2.9) has been used. Replacing the index P by L gives Eq. (2.2.16).

The derivative of the Jacobian with respect to one of its elements is

$$\frac{\partial J}{\partial y_{i,L}} = JX_{L,i}. \tag{2.2.18}$$

Proof. From Eq. (2.2.6)

$$6\frac{\partial J}{\partial y_{p,Q}} = \varepsilon_{ijk}\varepsilon_{LMN}\delta_{ip}\delta_{LQ}y_{j,M}y_{k,N} + \varepsilon_{ijk}\varepsilon_{LMN}y_{i,L}\delta_{jp}\delta_{MQ}y_{k,N}$$

$$+ \varepsilon_{ijk}\varepsilon_{LMN}y_{i,L}y_{j,M}\delta_{kp}\delta_{NQ}$$

$$= \varepsilon_{pjk}\varepsilon_{QMN}y_{j,M}y_{k,N} + \varepsilon_{ipk}\varepsilon_{LQN}y_{i,L}y_{k,N} + \varepsilon_{ijp}\varepsilon_{LMQ}y_{i,L}y_{j,M}$$

$$= 3\varepsilon_{pjk}\varepsilon_{QMN}y_{j,M}y_{k,N} = 3(2JX_{Q,p}), \tag{2.2.19}$$

where Eq. (2.2.16) has been used.

With Eq. (2.2.16), it can also be shown that

$$(JX_{K,k})_{,K} = 0. \tag{2.2.20}$$

Proof. The differentiation of both sides of Eq. (2.2.16) with respect to X_L gives

$$2(JX_{L,i})_{,L} = \varepsilon_{ijk}\varepsilon_{LMN}(y_{j,ML}y_{k,N} + y_{j,M}y_{k,NL}) = 0, \tag{2.2.21}$$

because

$$\varepsilon_{LMN}y_{j,ML} = 0, \quad \varepsilon_{LMN}y_{k,NL} = 0. \qquad (2.2.22)$$

Similarly, the following is true:

$$(J^{-1}y_{k,K})_{,k} = 0. \qquad (2.2.23)$$

2.3　Changes of Line, Area and Volume Elements

The lengths of a material line element before and after deformation are given by

$$(dL)^2 = dX_K dX_K = \delta_{KL} dX_K dX_L, \qquad (2.3.1)$$

and

$$\begin{aligned} (dl)^2 &= dy_i dy_i = y_{i,K} dX_K y_{i,L} dX_L \\ &= C_{KL} dX_K dX_L, \end{aligned} \qquad (2.3.2)$$

where C_{KL} is the deformation tensor defined by

$$C_{KL} = y_{k,K} y_{k,L} = C_{LK}. \qquad (2.3.3)$$

From Eqs. (2.3.1) and (2.3.2), we have

$$\begin{aligned} (dl)^2 - (dL)^2 &= (C_{KL} - \delta_{KL}) dX_K dX_L \\ &= 2E_{KL} dX_K dX_L, \end{aligned} \qquad (2.3.4)$$

where the finite strain tensor E_{KL} is defined by

$$\begin{aligned} E_{KL} &= (C_{KL} - \delta_{KL})/2 = (y_{i,K} y_{i,L} - \delta_{KL})/2 \\ &= (u_{K,L} + u_{L,K} + u_{M,K} u_{M,L})/2 = E_{LK}. \end{aligned} \qquad (2.3.5)$$

The unabbreviated form of Eq. (2.3.5) is

$$\begin{aligned} E_{11} &= u_{1,1} + (u_{1,1} u_{1,1} + u_{2,1} u_{2,1} + u_{3,1} u_{3,1})/2, \\ E_{22} &= u_{2,2} + (u_{1,2} u_{1,2} + u_{2,2} u_{2,2} + u_{3,2} u_{3,2})/2, \qquad (2.3.6) \\ E_{33} &= u_{3,3} + (u_{1,3} u_{1,3} + u_{2,3} u_{2,3} + u_{3,3} u_{3,3})/2, \end{aligned}$$

$$E_{23} = (u_{2,3} + u_{3,2} + u_{1,2}u_{1,3} + u_{2,2}u_{2,3} + u_{3,2}u_{3,3})/2,$$

$$E_{31} = (u_{3,1} + u_{1,3} + u_{1,3}u_{1,1} + u_{2,3}u_{2,1} + u_{3,3}u_{3,1})/2, \qquad (2.3.7)$$

$$E_{12} = (u_{1,2} + u_{2,1} + u_{1,1}u_{1,2} + u_{2,1}u_{2,2} + u_{3,1}u_{3,2})/2.$$

At the same material point, consider two material line elements $d\overset{(1)}{\mathbf{X}}$ and $d\overset{(2)}{\mathbf{X}}$ which deform into $d\overset{(1)}{\mathbf{y}}$ and $d\overset{(2)}{\mathbf{y}}$. The area of the parallelogram spanned by $d\overset{(1)}{\mathbf{X}}$ and $d\overset{(2)}{\mathbf{X}}$, and that by $d\overset{(1)}{\mathbf{y}}$ and $d\overset{(2)}{\mathbf{y}}$, can be represented by the following vectors, respectively:

$$N_L dS = dS_L = \varepsilon_{LMN} d\overset{(1)}{y}_M d\overset{(2)}{y}_N, \qquad (2.3.8)$$

$$n_i ds = ds_i = \varepsilon_{ijk} d\overset{(1)}{y}_j d\overset{(2)}{y}_k. \qquad (2.3.9)$$

They are related by

$$ds_i = J X_{L,i} dS_L. \qquad (2.3.10)$$

Proof.

$$ds_i = \varepsilon_{ijk} d\overset{(1)}{y}_j d\overset{(2)}{y}_k = \varepsilon_{ijk} y_{j,M} d\overset{(1)}{X}_M y_{k,N} d\overset{(2)}{X}_N$$

$$= \varepsilon_{ijk} y_{j,M} y_{k,N} d\overset{(1)}{X}_M d\overset{(2)}{X}_N = J X_{L,i} \varepsilon_{LMN} d\overset{(1)}{X}_M d\overset{(2)}{X}_N$$

$$= J X_{L,i} dS_L, \qquad (2.3.11)$$

where Eq. (2.2.12) has been used.

At the same material point, consider three material line elements $d\overset{(1)}{\mathbf{X}}$, $d\overset{(2)}{\mathbf{X}}$ and $d\overset{(3)}{\mathbf{X}}$ which deform into $d\overset{(1)}{\mathbf{y}}$, $d\overset{(2)}{\mathbf{y}}$ and $d\overset{(3)}{\mathbf{y}}$. The volume of the parallelepiped spanned by $d\overset{(1)}{\mathbf{X}}$, $d\overset{(2)}{\mathbf{X}}$ and $d\overset{(3)}{\mathbf{X}}$, and that by $d\overset{(1)}{\mathbf{y}}$,

$d\overset{(2)}{\mathbf{y}}$ and $d\overset{(3)}{\mathbf{y}}$, are related by

$$dv = JdV. \tag{2.3.12}$$

Proof.

$$
\begin{aligned}
dv &= d\overset{(1)}{\mathbf{y}} \cdot (d\overset{(2)}{\mathbf{y}} \times d\overset{(3)}{\mathbf{y}}) = \varepsilon_{ijk}d\overset{(1)}{y_i}d\overset{(2)}{y_j}d\overset{(3)}{y_k} \\
&= \varepsilon_{ijk}y_{i,L}d\overset{(1)}{X_L}y_{j,M}d\overset{(2)}{X_M}y_{k,N}d\overset{(3)}{X_N} \\
&= \varepsilon_{ijk}y_{i,L}y_{j,M}y_{k,N}d\overset{(1)}{X_L}d\overset{(2)}{X_M}d\overset{(3)}{X_N} \\
&= J\varepsilon_{LMN}d\overset{(1)}{X_L}d\overset{(2)}{X_M}d\overset{(3)}{X_N} \\
&= Jd\overset{(1)}{\mathbf{X}} \cdot (d\overset{(2)}{\mathbf{X}} \times d\overset{(3)}{\mathbf{X}}) = JdV, \tag{2.3.13}
\end{aligned}
$$

where Eq. (2.2.11) has been used.

2.4   Deformation Rates

Unless otherwise indicated, $\partial/\partial t$ is the time derivative with $\mathbf{y}$ fixed. The material time derivative with $\mathbf{X}$ fixed is denoted by d/dt or a superimposed dot. The convective time derivative is represented by an asterisk. The velocity and acceleration of a material point are given by the following material time derivatives:

$$v_i = \frac{dy_i}{dt} = \dot{y}_i = \left.\frac{\partial y_i(\mathbf{X}, t)}{\partial t}\right|_{\mathbf{X} \text{ fixed}}, \tag{2.4.1}$$

$$a_i = \frac{dv_i}{dt} = \dot{v}_i = \ddot{y}_i = \left.\frac{\partial^2 y_i(\mathbf{X}, t)}{\partial t^2}\right|_{\mathbf{X} \text{ fixed}}. \tag{2.4.2}$$

For a physical quantity in the form of $\phi[\mathbf{y}(\mathbf{X}, t), t]$, the material time derivative is given by

$$\frac{d\phi}{dt} = \dot{\phi} = \left.\frac{\partial\phi}{\partial t}\right|_{\mathbf{y} \text{ fixed}} + \left.\frac{\partial\phi}{\partial y_k}\frac{\partial y_k}{\partial t}\right|_{\mathbf{X} \text{ fixed}}$$

$$= \frac{\partial\phi}{\partial t} + v_k\frac{\partial\phi}{\partial y_k} = \frac{\partial\phi}{\partial t} + \mathbf{v}\cdot\nabla\phi, \tag{2.4.3}$$

where the spatial gradient operator is defined by

$$\nabla = \mathbf{i}_k\frac{\partial}{\partial y_k}. \tag{2.4.4}$$

The velocity gradient is defined by

$$\nabla\mathbf{v} = (\partial_k\mathbf{i}_k)(v_l\mathbf{i}_l) = v_{l,k}\mathbf{i}_k\mathbf{i}_l,$$
$$(\nabla\mathbf{v})_{kl} = v_{l,k}. \tag{2.4.5}$$

We also denote

$$\mathbf{v}\nabla = (v_k\mathbf{i}_k)(\partial_l\mathbf{i}_l) = v_{k,l}\mathbf{i}_k\mathbf{i}_l,$$
$$(\mathbf{v}\nabla)_{kl} = v_{k,l}. \tag{2.4.6}$$

The deformation rate tensor d_{ij} and the spin tensor ω_{ij} are introduced by decomposing the velocity gradient into symmetric and antisymmetric parts, i.e.,

$$v_{j,i} = d_{ij} + \omega_{ij},$$
$$\nabla\mathbf{v} = \mathbf{d} + \boldsymbol{\omega}, \tag{2.4.7}$$

where

$$d_{ij} = \frac{1}{2}(v_{j,i} + v_{i,j}), \quad \mathbf{d} = \frac{1}{2}(\nabla\mathbf{v} + \mathbf{v}\nabla), \tag{2.4.8}$$

$$\omega_{ij} = \frac{1}{2}(v_{j,i} - v_{i,j}), \quad \boldsymbol{\omega} = \frac{1}{2}(\nabla\mathbf{v} - \mathbf{v}\nabla). \tag{2.4.9}$$

We also have

$$\frac{d}{dt}(dy_i) = \frac{d}{dt}\left(\frac{\partial y_i}{\partial X_K}dX_K\right) = \frac{\partial}{\partial X_K}\left(\frac{dy_i}{dt}\right)dX_K$$

$$= \frac{\partial}{\partial X_K}(v_i)dX_K = v_{i,K}dX_K = v_{i,j}y_{j,K}dX_K. \quad (2.4.10)$$

The strain rate and the deformation rate are related by

$$\dot{E}_{KL} = d_{ij}y_{i,K}y_{j,L}. \quad (2.4.11)$$

Proof.

$$\dot{E}_{KL} = \frac{1}{2}(\dot{y}_{i,K}y_{i,L} + y_{i,K}\dot{y}_{i,L}) = \frac{1}{2}(v_{i,K}y_{i,L} + y_{i,K}v_{i,L})$$

$$= \frac{1}{2}(v_{i,j}y_{j,K}y_{i,L} + y_{i,K}v_{i,j}y_{j,L})$$

$$= \frac{1}{2}(v_{j,i}y_{i,K}y_{j,L} + y_{i,K}v_{i,j}y_{j,L})$$

$$= \frac{1}{2}(v_{j,i} + v_{i,j})y_{i,K}y_{j,L} = d_{ij}y_{i,K}y_{j,L}. \quad (2.4.12)$$

The material derivative of the Jacobian is given by

$$\dot{J} = Jv_{k,k}. \quad (2.4.13)$$

Proof. From Eq. (2.2.6),

$$\dot{J} = \frac{1}{6}\varepsilon_{klm}\varepsilon_{KLM}(v_{k,K}y_{l,L}y_{m,M} + y_{k,K}v_{l,L}y_{m,M} + y_{k,K}y_{l,L}v_{m,M})$$

$$= \frac{1}{2}\varepsilon_{klm}\varepsilon_{KLM}v_{k,K}y_{l,L}y_{m,M} = \frac{1}{2}v_{k,K}\varepsilon_{klm}\varepsilon_{KLM}y_{l,L}y_{m,M}$$

$$= \frac{1}{2}v_{k,K}2JX_{K,k} = Jv_{k,k}, \quad (2.4.14)$$

where Eq. (2.2.16) has been used.

The following expression is also useful:

$$\frac{d}{dt}(dv) = v_{k,k}dv. \tag{2.4.15}$$

Proof.

$$\frac{d}{dt}(dv) = \frac{d}{dt}(JdV) = \frac{dJ}{dt}dV$$
$$= Jv_{k,k}dV = v_{k,k}dv, \tag{2.4.16}$$

where Eq. (2.4.13) has been used. In addition,

$$\frac{d}{dt}(X_{L,j}) = -v_{i,K}X_{K,j}X_{L,i} = -v_{i,j}X_{L,i}. \tag{2.4.17}$$

Proof. Since

$$y_{i,K}X_{K,j} = \delta_{ij}, \tag{2.4.18}$$

we have, upon taking the material time derivative of both sides,

$$\dot{y}_{i,K}X_{K,j} + y_{i,K}\frac{d}{dt}(X_{K,j}) = 0. \tag{2.4.19}$$

Then

$$y_{i,K}\frac{d}{dt}(X_{K,j}) = -v_{i,K}X_{K,j}. \tag{2.4.20}$$

The multiplication of both sides of Eq. (2.4.20) by $X_{L,i}$ gives

$$\frac{d}{dt}(X_{L,j}) = -v_{i,K}X_{K,j}X_{L,i}. \tag{2.4.21}$$

In particular, for the material coordinates of a particle, $\mathbf{X} = \mathbf{X}(\mathbf{y}, t)$, we have $d\mathbf{X}/dt = 0$ or

$$\frac{\partial X_K}{\partial t} + v_k X_{K,k} = 0, \tag{2.4.22}$$

which leads to

$$v_k = -y_{k,K}\frac{\partial X_K}{\partial t}. \tag{2.4.23}$$

The following convective time derivative of a vector denoted by an asterisk is useful in continuum electrodynamics:

$$P_i^* = \dot{P}_i - v_{i,j}P_j + P_i v_{j,j}, \tag{2.4.24}$$

or

$$\mathbf{P}^* = \dot{\mathbf{P}} - (\mathbf{P} \cdot \nabla)\mathbf{v} + \mathbf{P}(\nabla \cdot \mathbf{v})$$

$$= \frac{\partial \mathbf{P}}{\partial t} + \nabla \times (\mathbf{P} \times \mathbf{v}) + \mathbf{v}(\nabla \cdot \mathbf{P}). \tag{2.4.25}$$

If

$$\mathcal{P}_K = JX_{K,i}P_i, \tag{2.4.26}$$

then

$$\dot{\mathcal{P}}_K = \left.\frac{\partial \mathcal{P}_K}{\partial t}\right|_{\mathbf{X}} = JX_{K,i}P_i^*. \tag{2.4.27}$$

2.5   Material Time Derivatives of Integrals

The material time derivative of a line integral is given by

$$\frac{d}{dt}\int_c \phi dy_k = \frac{d}{dt}\int_C \phi y_{k,L}dX_L = \int_C \frac{d}{dt}(\phi y_{k,L})dX_L$$

$$= \int_c (\dot{\phi}dy_k + \phi v_{k,l}dy_l). \tag{2.5.1}$$

The material time derivative of a surface integral is

$$\frac{d}{dt}\int_s \phi ds_k = \frac{d}{dt}\int_S \phi JX_{L,k}dS_L = \int_S \frac{d}{dt}\left(\phi JX_{L,k}\right)dS_L$$

$$= \int_s (\dot{\phi}ds_k + \phi v_{l,l}ds_k - \phi v_{l,k}ds_l). \tag{2.5.2}$$

A particular case of Eq. (2.5.2) is (when $\phi = q_k$):

$$\frac{d}{dt} \int_s q_k ds_k = \int_s (\dot{q}_k ds_k + q_k v_{l,l} ds_k - q_k v_{l,k} ds_l)$$

$$= \int_s (\dot{q}_k + q_k v_{l,l} - q_l v_{k,l}) ds_k$$

$$= \int_s \left(\frac{\partial q_k}{\partial t} + v_l q_{k,l} + q_k v_{l,l} - q_l v_{k,l} \right) ds_k$$

$$= \int_s q_k^* ds_k. \tag{2.5.3}$$

The material time derivative of a volume integral is

$$\frac{d}{dt} \int_v \phi dv = \frac{d}{dt} \int_V \phi J dV = \int_V \frac{d}{dt} (\phi J) dV$$

$$= \int_v (\dot{\phi} + \phi v_{k,k}) dv = \int_v \left[\frac{\partial \phi}{\partial t} + (\phi v_k)_{,k} \right] dv, \tag{2.5.4}$$

or

$$\frac{d}{dt} \int_v \phi dv = \int_v \frac{\partial \phi}{\partial t} dv + \int_s n_k v_k \phi ds, \tag{2.5.5}$$

which is usually referred to as the transport theorem.

2.6 Integrals Over a Changing Volume

In this section we discuss the time rate of change of an integral over an arbitrarily changing volume [13] whose boundary does not have to be a material surface. We begin with a surface moving and deforming in general defined by the following equation:

$$F(z_1, z_2, z_3, t) = 0. \tag{2.6.1}$$

On the surface,

$$\delta F = \frac{\partial F}{\partial t} \delta t + (\nabla F) \cdot \delta \mathbf{z} = 0,$$

$$\nabla F = \mathbf{e}_i \frac{\partial F}{\partial z_i}. \tag{2.6.2}$$

Then

$$\frac{\partial F}{\partial t} + |\nabla F| \frac{\nabla F}{|\nabla F|} \cdot \frac{\delta \mathbf{z}}{\delta t}$$

$$= \frac{\partial F}{\partial t} + |\nabla F| \mathbf{n} \cdot \mathbf{v}^s = \frac{\partial F}{\partial t} + |\nabla F| N = 0, \qquad (2.6.3)$$

where

$$\mathbf{n} = \frac{\nabla F}{|\nabla F|}, \qquad \mathbf{v}^s = \frac{\delta \mathbf{z}}{\delta t},$$

$$N = \mathbf{v}^s \cdot \mathbf{n} = -\frac{\partial F}{\partial t} \Big/ |\nabla F|. \qquad (2.6.4)$$

$\mathbf{n}$ is the unit normal of the surface at $\mathbf{z}$. $\mathbf{v}^s = \delta \mathbf{z}/\delta t$ is the velocity of the point of the surface at $\mathbf{z}$. N is the normal velocity of the point of the surface at $\mathbf{z}$.

When the surface is in a continuum whose velocity field is $\mathbf{v} = d\mathbf{y}/dt$, we denote the relative velocity of the surface with respect to the continuum at $\mathbf{y}$ in the direction of $\mathbf{n}$ by

$$\Theta = \mathbf{v}^s \cdot \mathbf{n} - \mathbf{v} \cdot \mathbf{n} = N - \mathbf{v} \cdot \mathbf{n} = N - v_k n_k$$

$$= -\frac{\partial F}{\partial t} \Big/ |\nabla F| - \mathbf{v} \cdot \frac{\nabla F}{|\nabla F|}$$

$$= \left(-\frac{\partial F}{\partial t} - v_k \frac{\partial F}{\partial y_k} \right) \Big/ |\nabla F| = -\frac{dF}{dt} \Big/ |\nabla F|. \qquad (2.6.5)$$

In the special case when the surface is a material surface consisting of the same particles of the continuum, $\mathbf{v}_s = \mathbf{v}$, $N = v_k n_k$ and Θ vanishes.

For the time rate of change of an integral over an arbitrarily changing volume v, to distinguish it from a material time derivative, we write

$$\frac{\delta}{\delta t} \int_{v(t)} \phi \, dv$$

$$= \lim_{\Delta t \to 0} \frac{1}{\Delta t} \left[\int_{v + \Delta v} \phi(\mathbf{y}, t + \Delta t) dv - \int_v \phi(\mathbf{y}, t) dv \right], \qquad (2.6.6)$$

which can be further written as

$$\frac{\delta}{\delta t}\int_{v(t)}\phi dv$$

$$= \lim_{\Delta t\to 0}\frac{1}{\Delta t}\left\{\int_v[\phi(\mathbf{y},t+\Delta t)-\phi(\mathbf{y},t)]dv\right.$$

$$\left.+\int_{\Delta v}\phi(\mathbf{y},t+\Delta t)dv\right\}, \qquad (2.6.7)$$

or

$$\frac{\delta}{\delta t}\int_{v(t)}\phi dv$$

$$= \int_v\frac{\partial\phi}{\partial t}dv + \lim_{\Delta t\to 0}\frac{1}{\Delta t}\int_{\Delta v}\phi(\mathbf{y},t+\Delta t)dv. \qquad (2.6.8)$$

Since

$$\int_{\Delta v}\phi dv = \int_{s(t)}\phi N\Delta t ds, \qquad (2.6.9)$$

where s is the boundary surface of v, we have

$$\frac{\delta}{\delta t}\int_{v(t)}\phi dv = \int_{v(t)}\frac{\partial\phi}{\partial t}dv + \int_{s(t)}\phi N ds. \qquad (2.6.10)$$

In the special case when s is a material surface, $\Theta = 0, N = v_k n_k$ and Eq. (2.6.10) reduces to

$$\frac{d}{dt}\int_v\phi dv = \int_v\frac{\partial\phi}{\partial t}dv + \int_s\phi n_k v_k ds, \qquad (2.6.11)$$

which is the transport theorem in Eq. (2.5.5).

2.7 A Material Body with a Discontinuity Surface

In this section, we discuss the time derivative of an integral over a material body occupying v containing a discontinuity surface Σ as shown in Fig. 2.2, and the related divergence theorem [10]. Similar

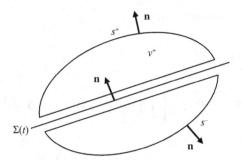

Fig. 2.2.　A material body with a discontinuity surface.

results about a material surface containing a discontinuity line can also be established [10]. Physically, Σ is a thin layer in which certain fields change rapidly across its thickness. Mathematically, Σ is treated as a surface across which certain fields are discontinuous and are not differentiable. Σ does not have to be material or planar. It divides the material body v into two parts. Although each part by itself is not a material body, the two parts together is. We denote

$$v = v^+ \cup v^-, \quad \partial v = s^+ \cup s^- = s,$$

$$\partial v^+ = s^+ \cup \Sigma, \quad \partial v^- = s^- \cup \Sigma. \tag{2.7.1}$$

Consider a field ϕ over v which may have a finite jump discontinuity over Σ and its derivative across Σ may be singular, like a delta function. We apply the general time derivative of a volume integral in Eq. (2.6.10) to the integration of ϕ over each part of v:

$$\frac{\delta}{\delta t} \int_{v^-} \phi dv = \int_{v^-} \frac{\partial \phi}{\partial t} dv + \int_{s^-} \phi v_k n_k ds + \int_{\Sigma} \phi^- N ds, \tag{2.7.2}$$

$$\frac{\delta}{\delta t} \int_{v^+} \phi dv = \int_{v^+} \frac{\partial \phi}{\partial t} dv + \int_{s^+} \phi v_k n_k ds - \int_{\Sigma} \phi^+ N ds. \tag{2.7.3}$$

Adding Eqs. (2.7.2) and (2.7.3), we obtain the time derivative of the integration of ϕ over the material body containing Σ as

$$\frac{d}{dt} \int_v \phi dv = \int_{v-\Sigma} \frac{\partial \phi}{\partial t} dv + \int_s \phi v_k n_k ds - \int_{\Sigma} [\phi] N ds, \tag{2.7.4}$$

where

$$[\phi] = \phi^+ - \phi^-, \tag{2.7.5}$$

and

$$\int_{v-\Sigma} (\ ) dv = \int_{v_1} (\ ) dv + \int_{v_2} (\ ) dv. \qquad (2.7.6)$$

The integration over $v - \Sigma$ is simply the sum of the integrations over v_1 and v_2. It does not have the contribution from Σ. In the special case when ϕ is continuous across Σ, we have $[\phi] = 0$ and Eq. (2.7.4) reduces to the transport theorem in Eq. (2.6.11) or (2.5.5).

Similarly, consider Φ over v containing a discontinuity surface Σ. Applying the conventional divergence theorem in Appendix C to each of the two parts of v in Fig. 2.2, we have

$$\int_{s^-} \mathbf{n} \cdot \mathbf{\Phi} ds + \int_{\Sigma} \mathbf{n} \cdot \mathbf{\Phi}^- ds = \int_{v^-} \nabla \cdot \mathbf{\Phi} dv, \qquad (2.7.7)$$

$$\int_{s^+} \mathbf{n} \cdot \mathbf{\Phi} ds + \int_{\Sigma} (-\mathbf{n}) \cdot \mathbf{\Phi}^+ ds = \int_{v^+} \nabla \cdot \mathbf{\Phi} dv. \qquad (2.7.8)$$

Adding Eqs. (2.7.7) and (2.7.8), we obtain the generalized divergence theorem below when a discontinuity surface is present:

$$\int_s \mathbf{n} \cdot \mathbf{\Phi} ds = \int_{v-\Sigma} \nabla \cdot \mathbf{\Phi} dv + \int_{\Sigma} \mathbf{n} \cdot [\mathbf{\Phi}] ds. \qquad (2.7.9)$$

When $\mathbf{n} \cdot \mathbf{\Phi}$ is continuous across Σ, Eq. (2.7.9) reduces to the conventional divergence theorem in Appendix C.

As a specific case, we let $\mathbf{\Phi} = \phi \mathbf{v}$ in Eq. (2.7.9). This yields

$$\int_s \phi v_k n_k ds = \int_{v-\Sigma} (\phi v_k)_{,k} dv + \int_{\Sigma} [\phi v_k n_k] ds. \qquad (2.7.10)$$

Substituting Eq. (2.7.10) into Eq. (2.7.4), we obtain

$$\frac{d}{dt} \int_v \phi dv = \int_{v-\Sigma} \left\{ \frac{\partial \phi}{\partial t} + (\phi v_k)_{,k} \right\} dv$$

$$+ \int_{\Sigma} [\phi(v_k n_k - N)] ds, \qquad (2.7.11)$$

or

$$\frac{d}{dt}\int_v \phi\, dv = \int_{v-\Sigma}\left\{\frac{\partial\phi}{\partial t} + (\phi v_k)_{,k}\right\} dv - \int_\Sigma [\phi\Theta]\, ds, \quad (2.7.12)$$

where, from Eq. (2.6.5),

$$\Theta = N - \mathbf{v}\cdot\mathbf{n}. \quad\quad\quad (2.7.13)$$

When $[\phi\Theta] = 0$, Eq. (2.7.12) reduces to Eq. (2.5.4).

Chapter 3

Elasticity

In this chapter, we develop the nonlinear theory of elasticity for large deformations. It is a special case of the nonlinear electromagnetoelastic theory to be developed in Chapters 4–6 and 8. Therefore this chapter serves as a preparation for the more involved chapters later. In addition to the basic theoretical framework, solutions to a few simple problems of elasticity are presented in this chapter. Some of them will be used for comparisons to solutions of similar problems with electromagnetic couplings in Chapter 7.

3.1 Balance Laws

For the theory of elasticity, the relevant laws from physics are the conservation of mass, linear momentum, angular momentum and energy as follows in integral form:

$$\frac{d}{dt}\int_v \rho \, dv = 0, \tag{3.1.1}$$

$$\frac{d}{dt}\int_v \rho \mathbf{v} \, dv = \int_v \rho \mathbf{f} \, dv + \int_s \mathbf{t} \, ds, \tag{3.1.2}$$

$$\frac{d}{dt}\int_v \mathbf{y} \times \rho \mathbf{v} \, dv = \int_v \mathbf{y} \times \rho \mathbf{f} \, dv + \int_s \mathbf{y} \times \mathbf{t} \, ds, \tag{3.1.3}$$

$$\frac{d}{dt}\int_v \rho \left(\frac{1}{2}\mathbf{v} \cdot \mathbf{v} + \varepsilon\right) dv = \int_v \rho \mathbf{f} \cdot \mathbf{v} \, dv + \int_s \mathbf{t} \cdot \mathbf{v} \, ds, \tag{3.1.4}$$

59

where ρ is the mass density, $\mathbf{f}$ the body force per unit mass, $\mathbf{t}$ the surface traction per unit area, and ε the internal energy density per unit mass. We convert Eqs. (3.1.1)–(3.1.4) into differential forms below.

Equation (3.1.1) states that the total mass of the material body is a constant, which is the total mass in the reference state, i.e.,

$$\int_v \rho dv = \int_V \rho^0 dV, \qquad (3.1.5)$$

where ρ^0 is the mass density in the reference state. For the left-hand side of Eq. (3.1.5), we apply the change of integration variables to the reference state using Eq. (2.3.12). Then the conservation of mass in Eq. (3.1.5) takes the following form:

$$\int_V (\rho J - \rho^0) dV = 0. \qquad (3.1.6)$$

Equation (3.1.6) holds for V or any part of it. Assuming a continuous integrand, we conclude from Eq. (3.1.6) that

$$\rho^0 = \rho J. \qquad (3.1.7)$$

Alternatively, since the total mass is conserved, we have

$$\frac{d}{dt} \int_v \rho dv = \frac{d}{dt} \int_V \rho J dV = \int_V \frac{d}{dt}(\rho J) dV$$

$$= \int_V (\dot{\rho} J + \rho \dot{J}) dV = \int_V (\dot{\rho} J + \rho J v_{i,i}) dV$$

$$= \int_v (\dot{\rho} + \rho v_{i,i}) dv = 0, \qquad (3.1.8)$$

where Eq. (2.4.13) has been used. Then

$$\dot{\rho} + \rho v_{i,i} = 0. \qquad (3.1.9)$$

From Eqs. (2.4.15) and (3.1.9), we can also obtain

$$\frac{d}{dt}(\rho dv) = \dot{\rho} dv + \rho \frac{d}{dt}(dv)$$

$$= -\rho v_{i,i} dv + \rho v_{i,i} dv = 0. \qquad (3.1.10)$$

Hence, for a differential material element of the body, the conservation of mass can be written as

$$\rho dv = \rho^0 dV. \tag{3.1.11}$$

With Eqs. (2.4.13) and (3.1.9), it can be shown that

$$\frac{d}{dt}\int_v \rho\phi dv = \int_v \rho\frac{d\phi}{dt}dv, \tag{3.1.12}$$

where ϕ represents a tensor field.

Proof. With the change of integration variables to the reference state, we have

$$
\begin{aligned}
\frac{d}{dt}\int_v \rho\phi dv &= \frac{d}{dt}\int_V \rho\phi J dV = \int_V \frac{d}{dt}(\rho\phi J)dV \\
&= \int_V \left(\dot{\rho}\phi J + \rho\frac{d\phi}{dt}J + \rho\phi\dot{J}\right)dV \\
&= \int_V \left(-\rho v_{i,i}\phi J + \rho\frac{d\phi}{dt}J + \rho\phi J v_{i,i}\right)dV \\
&= \int_V \rho\frac{d\phi}{dt}J dV = \int_v \rho\frac{d\phi}{dt}dv.
\end{aligned}\tag{3.1.13}
$$

For the conservation of linear momentum in Eq. (3.1.2), we introduce the Cauchy stress tensor τ by

$$t_i = n_j\tau_{ji} \tag{3.1.14}$$

through the tetrahedron argument [10]. Then, with the use of the divergence theorem, the balance of linear momentum becomes

$$
\begin{aligned}
\int_v \rho\frac{dv_i}{dt}dv &= \int_v \rho f_i dv + \int_s t_i ds \\
&= \int_v \rho f_i dv + \int_s \tau_{ji}n_j ds \\
&= \int_v \rho f_i dv + \int_v \tau_{ji,j}dv.
\end{aligned}\tag{3.1.15}
$$

Hence

$$\tau_{ji,j} + \rho f_i = \rho \dot{v}_i. \tag{3.1.16}$$

In terms of components, the balance of angular momentum in Eq. (3.1.3) takes the following form:

$$\frac{d}{dt} \int_v \varepsilon_{ijk} y_j \rho v_k dv = \int_v \varepsilon_{ijk} y_j \rho f_k dv + \int_s \varepsilon_{ijk} y_j t_k ds. \tag{3.1.17}$$

The term on the left-hand side of Eq. (3.1.17) can be written as

$$\frac{d}{dt} \int_v \varepsilon_{ijk} y_j \rho v_k dv = \int_v \rho \varepsilon_{ijk} \frac{d}{dt}(y_j v_k) dv$$

$$= \int_v \rho \varepsilon_{ijk} (\dot{y}_j v_k + y_j \dot{v}_k) dv$$

$$= \int_v \rho \varepsilon_{ijk} (v_j v_k + y_j \dot{v}_k) dv = \int_v \rho \varepsilon_{ijk} y_j \dot{v}_k dv. \tag{3.1.18}$$

The last term on the right-hand side of Eq. (3.1.17) can be written as

$$\int_s \varepsilon_{ijk} y_j t_k ds = \int_s \varepsilon_{ijk} y_j \tau_{lk} n_l ds$$

$$= \int_v (\varepsilon_{ijk} y_j \tau_{lk})_{,l} dv = \int_v \varepsilon_{ijk} (\delta_{jl} \tau_{lk} + y_j \tau_{lk,l}) dv$$

$$= \int_v \varepsilon_{ijk} (\tau_{jk} + y_j \tau_{lk,l}) dv. \tag{3.1.19}$$

Substituting Eqs. (3.1.18) and (3.1.19) into Eq. (3.1.17), we obtain

$$\int_v \rho \varepsilon_{ijk} y_j \dot{v}_k dv = \int_v \varepsilon_{ijk} y_j \rho f_k dv + \int_v \varepsilon_{ijk} (\tau_{jk} + y_j \tau_{lk,l}) dv, \tag{3.1.20}$$

or

$$\int_v \varepsilon_{ijk} y_j (\rho \ddot{v}_k - \rho f_k - \tau_{lk,l}) dv = \int_v \varepsilon_{ijk} \tau_{jk} dv. \qquad (3.1.21)$$

The left-hand side of Eq. (3.1.21) vanishes because of the linear momentum equation in Eq. (3.1.16). Hence,

$$\int_v \varepsilon_{ijk} \tau_{jk} dv = 0, \qquad (3.1.22)$$

which implies that

$$\varepsilon_{ijk} \tau_{jk} = 0. \qquad (3.1.23)$$

Equation (3.1.23) further implies that the Cauchy stress tensor τ_{kl} is symmetric, i.e., $\tau_{kl} = \tau_{lk}$.

In terms of components, the conservation of energy in Eq. (3.1.4) takes the following form:

$$\frac{d}{dt} \int_v \rho \left(\frac{1}{2} v_i v_i + \varepsilon \right) dv = \int_v \rho f_k v_k dv + \int_s t_k v_k ds. \qquad (3.1.24)$$

The left-hand side of Eq. (3.1.24) can be written as

$$\frac{d}{dt} \int_v \rho \left(\frac{1}{2} v_i v_i + \varepsilon \right) dv = \int_v \rho \frac{d}{dt} \left(\frac{1}{2} v_i v_i + \varepsilon \right) dv$$

$$= \int_v \rho (v_i \dot{v}_i + \dot{\varepsilon}) dv. \qquad (3.1.25)$$

The last term on the right-hand side of Eq. (3.1.24) can be written as

$$\int_s t_k v_k da = \int_s \tau_{lk} n_l v_k ds$$

$$= \int_v (\tau_{lk} v_k)_{,l} dv = \int_v (\tau_{lk,l} v_k + \tau_{lk} v_{k,l}) dv. \qquad (3.1.26)$$

The substitution of Eqs. (3.1.25) and (3.1.26) into Eq. (3.1.24) gives

$$\int_v \rho (v_k \dot{v}_k + \dot{\varepsilon}) dv = \int_v f_k v_k dv + \int_v (\tau_{lk,l} v_k + \tau_{lk} v_{k,l}) dv,$$

$$(3.1.27)$$

or

$$\int_v v_k(\rho\dot{v}_k - \rho f_k - \tau_{lk,l})dv = \int_v (\tau_{lk}v_{k,l} - \rho\dot{\varepsilon})dv. \tag{3.1.28}$$

The left-hand side of Eq. (3.1.28) vanishes because of the linear momentum equation in Eq. (3.1.16). Hence

$$\int_v (\tau_{lk}v_{k,l} - \rho\dot{\varepsilon})dv = 0, \tag{3.1.29}$$

which implies that

$$\rho\dot{\varepsilon} = \tau_{ij}v_{j,i}. \tag{3.1.30}$$

3.2   Constitutive Relations

For convenience, we gather the differential forms of the balance laws from Eqs. (3.1.9), (3.1.16), (3.1.23) and (3.1.30) as follows:

$$\dot{\rho} + \rho v_{i,i} = 0, \tag{3.2.1}$$

$$\rho\dot{v}_k = \tau_{ik,i} + \rho f_k, \tag{3.2.2}$$

$$\varepsilon_{ijk}\tau_{jk} = 0, \tag{3.2.3}$$

$$\rho\dot{\varepsilon} = \tau_{mk}v_{k,m}. \tag{3.2.4}$$

The above equations are written in terms of the present coordinates y_i in the sense that all spatial derivatives are taken with respect to **y**. Since the reference coordinates of the material points are known while the present coordinates are not, it is essential to have the equations written in terms of the reference coordinates X_K. For this purpose we introduce the first Piola–Kirchhoff stress tensor K_{Lj} which is a two-point tensor and the second Piola–Kirchhoff stress tensor P_{KL}

through

$$K_{Lj} = JX_{L,i}\tau_{ij}, \quad \tau_{ij} = J^{-1}y_{i,L}K_{Lj}, \tag{3.2.5}$$

$$P_{KL} = JX_{K,i}X_{L,j}\tau_{ij}, \quad \tau_{ij} = J^{-1}y_{i,K}y_{j,L}P_{KL}. \tag{3.2.6}$$

With K_{Lj}, we have

$$\int_s t_j ds = \int_s n_i \tau_{ij} ds = \int_s \tau_{ij} ds_i$$

$$= \int_S \tau_{ij} JX_{L,i} dS_L = \int_S \tau_{ij} JX_{L,i} N_L dS$$

$$= \int_V (\tau_{ij} JX_{L,i})_{,L} dV = \int_V K_{Lj,L} dV. \tag{3.2.7}$$

Then the linear momentum equation in Eq. (3.1.2) can be written as

$$\int_V \rho \dot{v}_j J dV = \int_V \rho f_j J dV + \int_V K_{Lj,L} dV, \tag{3.2.8}$$

which implies that

$$K_{Lj,L} + \rho^0 f_j = \rho^0 \dot{v}_j. \tag{3.2.9}$$

The angular momentum equation does not have a spatial derivative. In fact, it will be automatically satisfied later when the τ_{kl} given by the constitutive relation below is a symmetric tensor.

For the conservation of energy in Eq. (3.2.4), we have

$$\rho \dot{\varepsilon} = \tau_{ij} v_{j,i} = \tau_{ij}(d_{ij} + \omega_{ij}) = \tau_{ij} d_{ij}$$

$$= J^{-1} y_{i,K} y_{j,L} P_{KL} d_{ij} = J^{-1} P_{KL} \dot{E}_{KL}, \tag{3.2.10}$$

where Eq. (2.4.12) has been used. In summary, the balance laws in material form are

$$\rho^0 = \rho J, \tag{3.2.11}$$

$$K_{Lk,L} + \rho^0 f_k = \rho^0 \dot{v}_k, \tag{3.2.12}$$

$$\varepsilon_{kij}\tau_{ij} = 0, \tag{3.2.13}$$

$$\rho^0 \dot{\varepsilon} = P_{KL} \dot{E}_{KL}, \tag{3.2.14}$$

where Eq. (3.1.7) has been included and the spatial derivatives are taken with respect to **X**.

The above field equations represent the relevant basic laws of physics valid for a continuum in general. The specific behavior of a particular material is specified by additional equations called constitutive relations. For the constitutive relations of an elastic solid, we begin with the following internal energy density as suggested by Eq. (3.2.14):

$$\varepsilon = \varepsilon(E_{KL}). \tag{3.2.15}$$

The substitution of Eq. (3.2.15) into Eq. (3.2.14) gives

$$\left(P_{KL} - \rho^0 \frac{\partial \varepsilon}{\partial E_{KL}} \right) \dot{E}_{KL} = 0. \tag{3.2.16}$$

Since Eq. (3.2.16) is linear in $\dot{E}_{KL}$, for the equation to hold for any $\dot{E}_{KL} = \dot{E}_{LK}$, we must have

$$P_{KL} = \rho^0 \frac{1}{2} \left(\frac{\partial \varepsilon}{\partial E_{KL}} + \frac{\partial \varepsilon}{\partial E_{LK}} \right) = \rho^0 \frac{\partial \varepsilon}{\partial E_{KL}}, \tag{3.2.17}$$

where the partial derivatives with respect to E_{KL} are taken as if the strain components were independent, and ε is written as a symmetric function of E_{KL}, i.e.,

$$\frac{\partial E_{KL}}{\partial E_{LK}} = 0, \quad K \neq L,$$

$$\frac{\partial \varepsilon}{\partial E_{KL}} = \frac{\partial \varepsilon}{\partial E_{LK}}. \tag{3.2.18}$$

Then,

$$K_{Lj} = JX_{L,i}\tau_{ij} = JX_{L,i}(J^{-1}y_{i,K}y_{j,M}P_{KM})$$

$$= y_{j,M}P_{ML} = y_{j,K}\rho^0 \frac{\partial \varepsilon}{\partial E_{KL}}. \tag{3.2.19}$$

We also have

$$\tau_{ij} = J^{-1}y_{i,K}y_{j,L}\rho^0 \frac{\partial \varepsilon}{\partial E_{KL}} = \tau_{ji}. \tag{3.2.20}$$

The substitution Eq. (3.2.19) into Eq. (3.2.12) yields three equations for the three components of $\mathbf{y}(\mathbf{X}, t)$. Once $\mathbf{y}(\mathbf{X}, t)$ is known, the present mass density ρ can be obtained from the conservation of mass in Eq. (3.2.11).

3.3 Boundary Conditions

The field equations in Eqs. (3.2.1)–(3.2.4) in differential forms were obtained by applying the integral forms of the balance laws in Eqs. (3.1.1)–(3.1.4) to a material body in which the fields are continuous and differentiable. When there is a discontinuity surface in the material body, the implications of the integral balances laws need to be re-examined. Before we apply the integral balance laws in Eqs. (3.1.1)–(3.1.4) to a material body with a discontinuity surface, we summarize them into the following unified form:

$$\frac{d}{dt}\int_v \rho\Psi dv = \int_v \rho\Gamma dv + \int_s \mathbf{n}\cdot\mathbf{\Phi}ds. \tag{3.3.1}$$

The conservation of mass, linear momentum, angular momentum and energy correspond to

	Ψ	Γ	$\mathbf{\Phi}$	
Mass	1	0	0	
Linear momentum	$\mathbf{v}$	$\mathbf{f}$	$\mathbf{\tau}$	(3.3.2)
Angular momentum	$\mathbf{y}\times\mathbf{v}$	$\mathbf{y}\times\mathbf{f}$	$-\mathbf{\tau}\times\mathbf{y}$	
Energy	$\varepsilon+\mathbf{v}\cdot\mathbf{v}/2$	$\mathbf{f}\cdot\mathbf{v}$	$\mathbf{\tau}\cdot\mathbf{v}$	

When the volume v does not have a discontinuity surface, using the conventional transport theorem in Eq. (2.5.4) and conventional divergence theorem in Appendix C, the differential form of Eq. (3.3.1) is

$$\rho\frac{d\Psi}{dt} = \rho\Gamma + \nabla\cdot\mathbf{\Phi}. \tag{3.3.3}$$

When a discontinuity surface Σ is present in v as shown in Fig. 2.2, we need to use the modified transport theorem in Eq. (2.7.12) and the modified divergence theorem in Eq. (2.7.9). From Eq. (2.7.12), with $\phi = \rho\Psi$, we have

$$\frac{d}{dt}\int_v \rho\Psi dv = \int_{v-\Sigma}\left\{\frac{\partial(\rho\Psi)}{\partial t} + (\rho\Psi v_k)_{,k}\right\}dv - \int_\Sigma[\rho\Psi\Theta]ds.$$

$$(3.3.4)$$

From the generalized divergence theorem in Eq. (2.7.9),

$$\int_s \mathbf{n}\cdot\mathbf{\Phi}ds = \int_{v-\Sigma}\nabla\cdot\mathbf{\Phi}dv + \int_\Sigma \mathbf{n}\cdot[\mathbf{\Phi}]ds. \qquad (3.3.5)$$

The substitutions of Eqs. (3.3.4) and (3.3.5) into the general balance law in Eq. (3.3.1) gives

$$\int_{v-\Sigma}\left\{\frac{\partial(\rho\Psi)}{\partial t} + (\rho\Psi v_k)_{,k}\right\}dv - \int_\Sigma[\rho\Psi\Theta]ds$$

$$= \int_v \rho\Gamma dv + \int_{v-\Sigma}\nabla\cdot\mathbf{\Phi}dv + \int_\Sigma \mathbf{n}\cdot[\mathbf{\Phi}]ds, \qquad (3.3.6)$$

or

$$\int_{v-\Sigma}\left\{\frac{\partial(\rho\Psi)}{\partial t} + (\rho\Psi v_k)_{,k} - \rho\Gamma - \nabla\cdot\mathbf{\Phi}\right\}dv$$

$$= \int_\Sigma[\mathbf{n}\cdot\mathbf{\Phi} + \rho\Psi\Theta]ds, \qquad (3.3.7)$$

where $\rho\Gamma$ is assumed to be bounded in v so that its integration over v is the same as its integration over $v-\Sigma$. With the conservation of mass in Eq. (3.1.9) valid in v_1 and v_2, the two parts of v in Fig. 2.2, we can write Eq. (3.3.7) as

$$\int_{v-\Sigma}\left\{\rho\dot{\Psi} - \rho\Gamma - \nabla\cdot\mathbf{\Phi}\right\}dv = \int_\Sigma[\mathbf{n}\cdot\mathbf{\Phi} + \rho\Psi\Theta]ds. \qquad (3.3.8)$$

Equation (3.3.8) leads to Eq. (3.3.3) in $v-\Sigma$ or each of v_1 and v_2 as well as

$$\int_\Sigma[\rho\Psi\Theta + \mathbf{n}\cdot\mathbf{\Phi}]ds = 0, \qquad (3.3.9)$$

where $\Theta = N - \mathbf{v} \cdot \mathbf{n}$ (see Eq. (2.6.5)). Equation (3.3.9) implies the following jump condition across Σ in general:

$$[\rho \Psi \Theta + \mathbf{n} \cdot \boldsymbol{\Phi}] = 0. \tag{3.3.10}$$

From Eqs. (3.3.10) and (3.3.2) we have, for the conservation of mass, linear momentum, angular momentum and energy, the following jump conditions:

$$[\rho \Theta] = 0, \tag{3.3.11}$$

$$[\rho \mathbf{v} \Theta + \mathbf{n} \cdot \boldsymbol{\tau}] = 0, \tag{3.3.12}$$

$$[\rho \mathbf{y} \times \mathbf{v} \Theta + \mathbf{n} \cdot (-\boldsymbol{\tau} \times \mathbf{y})] = 0, \tag{3.3.13}$$

$$[\rho(\varepsilon + \mathbf{v} \cdot \mathbf{v}/2)\Theta + \mathbf{n} \cdot \boldsymbol{\tau} \cdot \mathbf{v} = 0. \tag{3.3.14}$$

We assume the continuity of $\mathbf{y}$ (and hence the continuity of $\mathbf{v}$) across the discontinuity surface. Then Eq. (3.3.13) is implied by Eq. (3.3.12).

In the special case when the discontinuity surface is material, $\Theta = 0$ and Eq. (3.3.11) becomes trivial. In this case, Eqs. (3.3.12) and (3.3.14) reduce to

$$[\mathbf{n} \cdot \boldsymbol{\tau}] = 0, \tag{3.3.15}$$

$$[\mathbf{n} \cdot \boldsymbol{\tau} \cdot \mathbf{v}] = 0. \tag{3.3.16}$$

Since $\mathbf{v}$ is continuous across the discontinuity surface, Eq. (3.3.15) implies Eq. (3.3.16).

If a material interface is under a distributed traction $\bar{t}_j$ per unit area of the interface, Eq. (3.3.15) needs to be modified. In this case the integral form of the linear momentum equation as a special case of Eq. (3.3.1) may still be considered as valid, but effectively, corresponding to $\bar{t}_j$, the body source Γ in Eq. (3.3.1) has a delta function type singularity at the interface. When applied to the pillbox in Fig. 3.1,

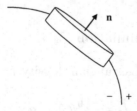

Fig. 3.1.   A pillbox on a discontinuity surface.

the integral form of the linear momentum equation implies that

$$\tau_{ij}^+ n_i - \tau_{ij}^- n_i + \bar{t}_j = 0. \tag{3.3.17}$$

Since

$$\tau_{ij} n_i ds = \tau_{ij} ds_i = \tau_{ij} J X_{L,i} dS_L$$
$$= K_{Lj} dS_L = K_{Lj} N_L dS, \tag{3.3.18}$$

Eq. (3.3.17) can be written in the material form as

$$N_L K_{Lj}^+ - N_L K_{Lj}^- + \bar{T}_j = 0, \tag{3.3.19}$$

where

$$\bar{T}_j = \bar{t}_j \frac{ds}{dS}. \tag{3.3.20}$$

Consider a finite body V whose boundary surface S is partitioned into S_y and S_T on which position and traction are prescribed, respectively. We assume that

$$S_y \cup S_T = S, \quad S_y \cap S_T = \emptyset. \tag{3.3.21}$$

Usual boundary-value problems consist of Eqs. (3.2.12) and (3.2.19) with the following boundary conditions:

$$y_i = \bar{y}_i \quad \text{on} \quad S_y,$$
$$N_L K_{Lk} = \bar{T}_k \quad \text{on} \quad S_T, \tag{3.3.22}$$

where $\bar{y}_i$ is prescribed boundary position, and $\bar{T}_i$ is prescribed surface traction per unit undeformed area. For dynamic problems, initial conditions need to be added.

3.4   Variational Formulation

Consider the following Lagrangian density L:

$$L = \frac{1}{2} \rho^0 \dot{y}_i \dot{y}_i - \rho^0 \varepsilon \tag{3.4.1}$$

and variational functional Π:

$$\Pi(\mathbf{y}) = \int_{t_0}^{t_1} dt \int_V [L + \rho^0 f_i y_i] dV + \int_{t_0}^{t_1} dt \int_{S_T} \bar{T}_i y_i dS,$$

$$= \int_{t_0}^{t_1} dt \int_V \left[\frac{1}{2} \rho^0 \dot{y}_i \dot{y}_i - \rho^0 \varepsilon(\mathbf{E}) + \rho^0 f_i y_i \right] dV$$

$$+ \int_{t_0}^{t_1} dt \int_{S_T} \bar{T}_i y_i dS, \qquad (3.4.2)$$

where

$$E_{KL} = (y_{i,K} y_{i,L} - \delta_{KL})/2. \qquad (3.4.3)$$

The admissible $\mathbf{y}$ for Π satisfies the following conditions at t_0 and t_1 as well as the essential boundary conditions on S_y:

$$\delta y_i|_{t=t_0} = 0, \quad \delta y_i|_{t=t_1} = 0 \quad \text{in} \quad V,$$

$$y_i = \bar{y}_i \quad \text{on} \quad S_y, \quad t_0 < t < t_1. \qquad (3.4.4)$$

Then the first variation of Π can be found as

$$\delta\Pi = \int_{t_0}^{t_1} dt \int_V (K_{Li,L} + \rho^0 f_i - \rho^0 \ddot{y}_i) \delta y_i \, dV$$

$$- \int_{t_0}^{t_1} dt \int_{S_T} (N_L K_{Li} - \bar{T}_i) \delta y_i dS, \qquad (3.4.5)$$

where

$$K_{Lj} = y_{j,K} \rho^0 \frac{\partial \varepsilon}{\partial E_{KL}}. \qquad (3.4.6)$$

Therefore, the stationary condition of Π implies the following field equations and natural boundary conditions:

$$K_{Lk,L} + \rho^0 f_k = \rho^0 \ddot{y}_k \quad \text{in} \quad V,$$

$$N_L K_{Lk} = \bar{T}_k \quad \text{on} \quad S_T. \qquad (3.4.7)$$

3.5 Third-Order Theory

The internal energy density ε that determines the constitutive relations of nonlinear elastic materials may be written as [12]

$$\rho^0 \varepsilon(\mathbf{E}) = \frac{1}{2} \underset{2}{c}_{ABCD} E_{AB} E_{CD} + \frac{1}{6} \underset{3}{c}_{ABCDEF} E_{AB} E_{CD} E_{EF}$$

$$+ \frac{1}{24} \underset{4}{c}_{ABCDEFGH} E_{AB} E_{CD} E_{EF} E_{GH} + \cdots, \qquad (3.5.1)$$

where a constant term and a term linear in $\mathbf{E}$ describing initial stress have been dropped. For weak nonlinearity, the higher-order terms in Eq. (3.5.1) may be neglected. The material constants in Eq. (3.5.1),

$$\underset{2}{c}_{ABCD}, \quad \underset{3}{c}_{ABCDEF}, \quad \underset{4}{c}_{ABCDEFGH}, \qquad (3.5.2)$$

are the second-, third- and fourth-order fundamental elastic constants, respectively. The second-order constants are mainly responsible for linear material behaviors. The third- and higher-order material constants are for nonlinear behaviors. The structure of $\varepsilon(\mathbf{E})$ depends on material symmetry.

By a third-order theory, we mean that the effects of all terms up to the third power of the displacement gradient components or their products are kept, but higher-order terms are neglected. Such a theory can be obtained by truncations of the equations of the fully nonlinear theory [12]. This results in the following expression for K_{Lj}:

$$K_{Lj} \cong \delta_{jM} \left[\underset{2}{c}_{LMAB} u_{A,B} + \frac{1}{2} \underset{2}{c}_{LMAB} u_{K,A} u_{K,B} + \underset{2}{c}_{LKAB} u_{M,K} u_{A,B} \right.$$

$$+ \frac{1}{2} \underset{3}{c}_{LMABCD} u_{A,B} u_{C,D} + \frac{1}{2} \underset{2}{c}_{LRAB} u_{M,R} u_{K,A} u_{K,B}$$

$$+ \frac{1}{2} \underset{3}{c}_{LKABCD} u_{M,K} u_{A,B} u_{C,D}$$

$$+ \frac{1}{2} \underset{3}{c}_{LMABCD} u_{A,B} u_{K,C} u_{K,D}$$

$$\left. + \frac{1}{6} \underset{4}{c}_{LMABCDEF} u_{A,B} u_{C,D} u_{E,F} \right]. \qquad (3.5.3)$$

We note that the fourth-order material constants are needed for a complete description of the third-order effects of the small displacement gradient.

As a special case of the third-order theory, if we keep terms up to the second order of the small displacement gradients only, we obtain the second-order theory as

$$K_{Lj} \cong \delta_{jM} \left[\underset{2}{c}_{LMAB} u_{A,B} + \frac{1}{2} \underset{2}{c}_{LMAB} u_{K,A} u_{K,B} \right.$$

$$\left. + \underset{2}{c}_{LKAB} u_{M,K} u_{A,B} + \frac{1}{2} \underset{3}{c}_{LMABCD} u_{A,B} u_{C,D} \right]. \quad (3.5.4)$$

The first-order or the linear constitutive relation is given by

$$K_{Lj} \cong \delta_{jM} \underset{2}{c}_{LMAB} u_{A,B}. \quad (3.5.5)$$

3.6 Linear Theory for Small Deformations

The linear theory of elasticity is for small-amplitude motions of an elastic body around its reference state under small mechanical loads. It is assumed that under some norm the displacement gradient is infinitesimal, e.g.,

$$\|u_{i,K}\| \ll 1, \quad \|u_{i,K}\| = \max |u_{i,K}|. \quad (3.6.1)$$

We want to obtain equations linear in $u_{i,M}$ and neglect their products or powers. For example,

$$\frac{\partial u_i}{\partial X_K} = \frac{\partial u_i}{\partial y_k} y_{k,K} = \frac{\partial u_i}{\partial y_k} (\delta_{kK} + u_{k,K}) \cong \frac{\partial u_i}{\partial y_k} \delta_{kK}, \quad (3.6.2)$$

which implies that, to the first order of approximation, the small displacement gradients calculated from the material and spatial coordinates are approximately equal. The material time derivative of an infinitesimal field $f(\mathbf{y}, t) = f[\mathbf{y}(\mathbf{X}, t), t]$ is approximately equal to:

$$\frac{df}{dt} = \frac{\partial f}{\partial t}\bigg|_{\mathbf{X} \text{ fixed}} = \frac{\partial f}{\partial t}\bigg|_{\mathbf{y} \text{ fixed}} + \frac{\partial f}{\partial y_i}\bigg|_{t \text{ fixed}} \frac{\partial y_i}{\partial t}\bigg|_{\mathbf{X} \text{ fixed}}$$

$$= \frac{\partial f}{\partial t}\bigg|_{\mathbf{y} \text{ fixed}} + \frac{\partial f}{\partial y_i}\bigg|_{t \text{ fixed}} v_i \cong \frac{\partial f}{\partial t}\bigg|_{\mathbf{y} \text{ fixed}}. \qquad (3.6.3)$$

For the finite strain tensor, we have, approximately,

$$E_{KL} = \frac{1}{2}(u_{L,K} + u_{K,L} + u_{M,K}u_{M,L}) \cong \frac{1}{2}(u_{L,K} + u_{K,L}). \qquad (3.6.4)$$

In linear elasticity, the capital $\mathbf{X}$ is usually replaced by the lowercase $\mathbf{x}$, and only lowercase indices are used. The infinitesimal strain tensor is denoted by [14]

$$S_{kl} = \frac{1}{2}(u_{l,k} + u_{k,l}) = S_{lk}. \qquad (3.6.5)$$

Similarly, we have, for the small stresses,

$$K_{Lj} \cong \delta_{Li}\tau_{ij}, \quad P_{KL} \cong \delta_{Ki}\delta_{Lj}\tau_{ij}. \qquad (3.6.6)$$

Since the stress tensors are approximately the same in the linear theory, we use T_{ij} to denote the infinitesimal stress tensor [14], i.e.,

$$\tau_{ij} \cong \to T_{ij},$$

$$K_{Lj} \cong \delta_{Li}\tau_{ij} \to T_{lj}, \qquad (3.6.7)$$

$$P_{KL} \cong \delta_{Ki}\delta_{Lj}\tau_{ij} \to T_{kl}.$$

We also introduce the internal energy density U per unit volume as follows:

$$U(\mathbf{E}) = \rho^0 \varepsilon(\mathbf{E}) \cong \frac{1}{2} \underset{2}{c}_{ABCD} E_{AB} E_{CD}$$

$$\to \frac{1}{2} c_{ijkl} S_{ij} S_{kl}. \qquad (3.6.8)$$

The linear constitutive relation generated by U is

$$T_{ij} = \frac{\partial U}{\partial S_{ij}} = c_{ijkl}S_{kl}. \tag{3.6.9}$$

The elastic stiffness c_{ijkl} has the following symmetries:

$$c_{ijkl} = c_{jikl} = c_{klij}. \tag{3.6.10}$$

We also assume that the elastic stiffness is positive definite in the following sense:

$$
\begin{aligned}
c_{ijkl}S_{ij}S_{kl} \geq 0 \quad &\text{for any} \quad S_{ij} = S_{ji}, \\
\text{and} \quad c_{ijkl}S_{ij}S_{kl} = 0 \quad &\Rightarrow \quad S_{ij} = 0.
\end{aligned}
\tag{3.6.11}
$$

The linear constitutive relations in Eq. (3.6.9) can also be written as

$$S_{ij} = s_{ijkl}T_{kl}, \tag{3.6.12}$$

where s_{ijkl} is the elastic compliance.

In the notation of the linear theory of elasticity, the equations of motion is written as

$$T_{ji,j} + f_i = \rho\ddot{u}_i, \tag{3.6.13}$$

where **f** is the body force per unit volume and ρ the known mass density in the reference state which was written as ρ^0 in the nonlinear theory in the previous sections. The surface loads are also infinitesimal. We have

$$\bar{T}_j \to \bar{t}_j, \quad N_L K_{Lj} \to n_l T_{lj}. \tag{3.6.14}$$

We now introduce a compact matrix notation [14]. It allows us to use matrices to represent the stress, strain, elastic stiffness and compliance tensors and is convenient for constitutive relations of anisotropic materials. This notation consists of replacing pairs of tensor indices ij or kl by single matrix indices p or q, where i, j, k

and l take the values of 1, 2 and 3. p and q take the values of 1, 2, 3, 4, 5 and 6 according to

$$ij \text{ or } kl: \quad 11 \quad 22 \quad 33 \quad 23 \text{ or } 32 \quad 31 \text{ or } 13 \quad 12 \text{ or } 21$$
$$p \text{ or } q: \quad 1 \quad 2 \quad 3 \quad 4 \quad 5 \quad 6 \qquad (3.6.15)$$

Thus, for the stress tensor, we write

$$T_1 = T_{11}, \quad T_2 = T_{22}, \quad T_3 = T_{33},$$
$$T_4 = T_{23}, \quad T_5 = T_{31}, \quad T_6 = T_{12}. \qquad (3.6.16)$$

For the strain tensor, we introduce S_p such that

$$S_1 = S_{11}, \quad S_2 = S_{22}, \quad S_3 = S_{33},$$
$$S_4 = 2S_{23}, \quad S_5 = 2S_{31}, \quad S_6 = 2S_{12}. \qquad (3.6.17)$$

Accordingly, the strain energy density per unit volume can be written as

$$U(\boldsymbol{S}) = \frac{1}{2}T_pS_p = \frac{1}{2}c_{pq}S_pS_q. \qquad (3.6.18)$$

The corresponding constitutive relations take the following form:

$$T_p = \frac{\partial U}{\partial S_p} = c_{pq}S_q, \quad c_{pq} = c_{qp},$$
$$S_p = s_{pq}T_q, \quad s_{pq} = s_{qp}. \qquad (3.6.19)$$

The elastic properties of a fully anisotropic material such as a triclinic crystal are represented by the following full array of $[c_{pq}]$ with twenty one independent material constants:

$$
\begin{bmatrix} T_1 \\ T_2 \\ T_3 \\ T_4 \\ T_5 \\ T_6 \end{bmatrix}
=
\begin{bmatrix}
c_{11} & c_{12} & c_{13} & c_{14} & c_{15} & c_{16} \\
c_{21} & c_{22} & c_{23} & c_{24} & c_{25} & c_{26} \\
c_{31} & c_{32} & c_{33} & c_{34} & c_{35} & c_{36} \\
c_{41} & c_{42} & c_{43} & c_{44} & c_{45} & c_{46} \\
c_{51} & c_{52} & c_{53} & c_{54} & c_{55} & c_{56} \\
c_{61} & c_{62} & c_{63} & c_{64} & c_{65} & c_{66}
\end{bmatrix}
\begin{bmatrix} S_1 \\ S_2 \\ S_3 \\ S_4 \\ S_5 \\ S_6 \end{bmatrix}. \qquad (3.6.20)
$$

In the special case of an isotropic material with two independent material constants, the stress-strain relation reduces to

$$\begin{aligned}
T_1 &= c_{11}S_1 + c_{12}S_2 + c_{12}S_3, \\
T_2 &= c_{21}S_1 + c_{11}S_2 + c_{12}S_3, \\
T_3 &= c_{21}S_1 + c_{21}S_2 + c_{11}S_3, \\
T_4 &= c_{44}S_4, \quad T_5 = c_{44}S_5, \quad T_6 = c_{44}S_6,
\end{aligned}$$

(3.6.21)

where

$$c_{44} = \frac{1}{2}(c_{11} - c_{12}).$$

(3.6.22)

The relations of the constants c_{11} and c_{12} to Lamé's constants (λ, μ) and to Young's modulus, E, and Poisson's ratio, ν, are given in Table 3.1 [14].

For transversely isotropic materials and hexagonal crystals of class (6mm), their stiffness matrices have the same structure. Let the axis

Table 3.1. Elastic constants of isotropic materials.

	c_{11}, c_{12}	λ, μ	E, ν
c_{11}	c_{11}	$\lambda + 2\mu$	$\dfrac{E(1 - \nu)}{(1 + \nu)(1 - 2\nu)}$
c_{12}	c_{12}	λ	$\dfrac{E\nu}{(1 + \nu)(1 - 2\nu)}$
λ	c_{12}	λ	$\dfrac{E\nu}{(1 + \nu)(1 - 2\nu)}$
μ	$\dfrac{c_{11} - c_{12}}{2}$	μ	$\dfrac{E}{2(1 + \nu)}$
E	$\dfrac{(c_{11} + 2c_{12})(c_{11} - c_{12})}{c_{11} + c_{12}}$	$\dfrac{\mu(3\lambda + 2\mu)}{\lambda + \mu}$	E
ν	$\dfrac{c_{12}}{c_{11} + c_{12}}$	$\dfrac{\lambda}{2(\lambda + \mu)}$	ν

of rotational symmetry be x_3. The stiffness matrix is

$$
\begin{pmatrix}
c_{11} & c_{12} & c_{13} & 0 & 0 & 0 \\
c_{21} & c_{11} & c_{13} & 0 & 0 & 0 \\
c_{31} & c_{31} & c_{33} & 0 & 0 & 0 \\
0 & 0 & 0 & c_{44} & 0 & 0 \\
0 & 0 & 0 & 0 & c_{44} & 0 \\
0 & 0 & 0 & 0 & 0 & c_{66}
\end{pmatrix},
\qquad (3.6.23)
$$

where $c_{66} = (c_{11} - c_{12})/2$. The corresponding constitutive relations take the following form:

$$
\begin{aligned}
T_{11} &= c_{11}u_{1,1} + c_{12}u_{2,2} + c_{13}u_{3,3}, \\
T_{22} &= c_{12}u_{1,1} + c_{11}u_{2,2} + c_{13}u_{3,3}, \\
T_{33} &= c_{13}u_{1,1} + c_{13}u_{2,2} + c_{33}u_{3,3}, \\
\end{aligned}
\qquad (3.6.24)
$$

$$
\begin{aligned}
T_{23} &= c_{44}(u_{2,3} + u_{3,2}), \\
T_{31} &= c_{44}(u_{3,1} + u_{1,3}), \\
T_{12} &= c_{66}(u_{1,2} + u_{2,1}).
\end{aligned}
\qquad (3.6.25)
$$

The equations of motion are

$$
\begin{aligned}
&c_{11}u_{1,11} + (c_{12} + c_{66})u_{2,12} + (c_{13} + c_{44})u_{3,13} \\
&\quad + c_{66}u_{1,22} + c_{44}u_{1,33} + f_1 = \rho\ddot{u}_1,
\end{aligned}
\qquad (3.6.26)
$$

$$
\begin{aligned}
&c_{66}u_{2,11} + (c_{12} + c_{66})u_{1,12} + c_{11}u_{2,22} \\
&\quad + (c_{13} + c_{44})u_{3,23} + c_{44}u_{2,33} + f_2 = \rho\ddot{u}_2,
\end{aligned}
\qquad (3.6.27)
$$

$$
\begin{aligned}
&c_{44}u_{3,11} + (c_{44} + c_{13})u_{1,31} + c_{44}u_{3,22} \\
&\quad + (c_{13} + c_{44})u_{2,23} + c_{33}u_{3,33} + f_3 = \rho\ddot{u}_3.
\end{aligned}
\qquad (3.6.28)
$$

For fields and motions independent of x_3, Eqs. (3.6.26)–(3.6.28) reduce to two uncoupled systems:

$$c_{11}u_{1,11} + (c_{12} + c_{66})u_{2,12} + c_{66}u_{1,22} = \rho\ddot{u}_1,$$

$$c_{66}u_{2,11} + (c_{12} + c_{66})u_{1,12} + c_{11}u_{2,22} = \rho\ddot{u}_2, \tag{3.6.29}$$

$$c_{44}(u_{3,11} + u_{3,22}) = \rho\ddot{u}_3. \tag{3.6.30}$$

The motion described by u_3 in Eq. (3.6.30) is the so-called antiplane or shear-horizontal (SH) motion. Equation (3.6.29) describes plane-strain motions.

3.7   Plane Waves

For some simple problems of linear elasticity, consider plane waves in an unbounded cubic crystal of class (m3m) for which

$$[c_{pq}] = \begin{bmatrix} c_{11} & c_{12} & c_{12} & 0 & 0 & 0 \\ c_{12} & c_{11} & c_{12} & 0 & 0 & 0 \\ c_{12} & c_{12} & c_{11} & 0 & 0 & 0 \\ 0 & 0 & 0 & c_{44} & 0 & 0 \\ 0 & 0 & 0 & 0 & c_{44} & 0 \\ 0 & 0 & 0 & 0 & 0 & c_{44} \end{bmatrix}. \tag{3.7.1}$$

The stress-strain relation are

$$T_1 = c_{11}S_1 + c_{12}S_2 + c_{12}S_3,$$

$$T_2 = c_{12}S_1 + c_{11}S_2 + c_{12}S_3,$$

$$T_3 = c_{12}S_1 + c_{12}S_2 + c_{11}S_3, \tag{3.7.2}$$

$$T_4 = c_{44}S_4, \quad T_5 = c_{44}S_5, \quad T_6 = c_{44}S_6.$$

Consider waves propagating in the x_3 direction without x_1 and x_2 dependence. They are governed by

$$c_{44}u_{1,33} = \rho\ddot{u}_1,$$

$$c_{44}u_{2,33} = \rho\ddot{u}_2, \tag{3.7.3}$$

$$c_{11}u_{3,33} = \rho\ddot{u}_3.$$

In Eq. $(3.7.3)_1$, let

$$u_1 = U_1 \exp[i(\zeta x_3 - \omega t)], \qquad (3.7.4)$$

which is a transverse or shear wave. The substitution of Eq. (3.7.4) into Eq. $(3.7.3)_1$ leads to the following wave speed:

$$v = \frac{\omega}{\zeta} = \sqrt{\frac{c_{44}}{\rho}}. \qquad (3.7.5)$$

Equation $(3.7.3)_2$ also describes a shear wave with the same speed. For Eq. $(3.7.3)_3$, let

$$u_3 = U_3 \exp[i(\zeta x_3 - \omega t)], \qquad (3.7.6)$$

which is a longitudinal wave. The substitution of Eq. (3.7.6) into Eq. $(3.7.3)_3$ leads to the following wave speed:

$$v = \frac{\omega}{\zeta} = \sqrt{\frac{c_{11}}{\rho}}. \qquad (3.7.7)$$

3.8 Thickness Vibrations of Plates

Consider a plate of cubic crystals as shown in Fig. 3.2. It has a uniform thickness $2h$ and is unbounded in the x_1 and x_2 directions. The two surfaces are traction free. The boundary conditions are

$$T_{3j} = 0, \quad x_3 = \pm h. \qquad (3.8.1)$$

Thickness modes depend on the plate thickness coordinate x_3 and time only. For cubic crystals, thickness vibrations separate into thickness-shear and thickness-stretch modes. Consider

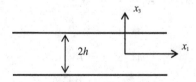

Fig. 3.2. An unbounded plate of cubic crystals.

thickness-shear modes described by u_1 first. They are governed by Eqs. $(3.7.3)_1$ and $(3.8.1)$:

$$c_{44}u_{1,33} = \rho\ddot{u}_1, \quad -h < x_3 < h,$$
$$T_{31} = T_5 = c_{44}u_{1,3} = 0, \quad x_3 = \pm h. \tag{3.8.2}$$

Consider modes antisymmetric about the plate middle plane first. Let

$$u_1 = U_1 \sin \zeta\, x_3 \exp(i\omega t). \tag{3.8.3}$$

The boundary conditions in Eq. $(3.8.2)_2$ imply that

$$\zeta^{(n)} = \frac{n\pi}{2h}, \quad n = 1, 3, 5, \ldots. \tag{3.8.4}$$

Equation $(3.8.2)_1$ requires that

$$\omega^{(n)} = \sqrt{\frac{c_{44}}{\rho}}\zeta^{(n)} = \sqrt{\frac{c_{44}}{\rho}}\frac{n\pi}{2h}, \quad n = 1, 3, 5, \ldots. \tag{3.8.5}$$

$n = 1$ gives the fundamental thickness-shear frequency and mode. $n > 1$ are the overtone frequencies and modes. Equation $(3.8.5)$ shows that the overtones are integral multiples of the fundamental and are called harmonic overtones or harmonics. If $\cos\zeta x_3$ is used in Eq. $(3.8.3)$, a different set of modes symmetric about $x_3 = 0$ with $n = 0, 2, 4, \ldots$ will be obtained. $n = 0$ represents a rigid-body mode and is not of interest. Variations of u_1 along the plate thickness for the static thickness-shear deformation and the first few thickness-shear modes are shown in Fig. 3.3. Thickness-shear modes described by u_2 are similar.

Thickness-stretch vibrations are governed by Eqs. $(3.7.3)_3$ and $(3.8.1)$:

$$c_{11}u_{3,33} = \rho\ddot{u}_3, \quad -h < x_3 < h,$$
$$T_{33} = T_3 = c_{11}u_{3,3} = 0, \quad x_3 = \pm h. \tag{3.8.6}$$

Let

$$u_3 = U_3 \sin \zeta\, x_3 \exp(i\omega t). \tag{3.8.7}$$

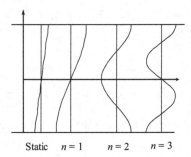

Fig. 3.3.   Variations of u_1 along the plate thickness for the static thickness-shear deformation and the first few thickness-shear modes.

The boundary conditions in Eq. $(3.8.6)_2$ imply that

$$\zeta^{(n)} = \frac{n\pi}{2h}, \quad n = 1, 3, 5, \ldots. \tag{3.8.8}$$

Equation $(3.8.6)_1$ requires that

$$\omega^{(n)} = \sqrt{\frac{c_{11}}{\rho}}\zeta^{(n)} = \sqrt{\frac{c_{11}}{\rho}}\frac{n\pi}{2h}, \quad n = 1, 3, 5, \ldots. \tag{3.8.9}$$

If $\cos\zeta x_3$ is used in Eq. (3.8.7), a different set of modes with $n = 0, 2, 4, \ldots$ will be obtained.

3.9   Waves in Plates

Consider the so-called antiplane or SH waves described by $u_1 = 0$, $u_2 = 0$ and $u_3 = u_3(x_1, x_2, t)$ in a plate of cubic crystals as shown in Fig. 3.4. The surfaces of the plate are traction free. The governing equation is

$$T_{13,1} + T_{23,2} = c_{44}(u_{3,11} + u_{3,22})$$

$$= c_{44}\nabla^2 u_3 = \rho\ddot{u}_3, \quad -h < x_2 < h. \tag{3.9.1}$$

The boundary conditions are

$$T_{23} = c_{44}u_{3,2} = 0, \quad x_2 = \pm h. \tag{3.9.2}$$

There are two types of waves that can propagate in the plate. One is symmetric and the other antisymmetric about the plate middle

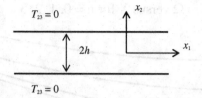

Fig. 3.4. An unbounded plate of cubic crystals.

plane. We discuss them separately below. For antisymmetric waves we consider the possibility of

$$u_3 = U_3 \sin \eta\, x_2 \cos(\xi x_1 - \omega t). \tag{3.9.3}$$

For Eq. (3.9.3) to satisfy Eq. (3.9.1), we must have

$$\eta^2 = \frac{\rho \omega^2}{c_{44}} - \xi^2 = \xi^2 \left(\frac{v^2}{v_T^2} - 1 \right),$$

$$v^2 = \frac{\omega^2}{\xi^2}, \quad v_T^2 = \frac{c_{44}}{\rho}. \tag{3.9.4}$$

The substitution of Eq. (3.9.3) into Eq. (3.9.2) leads to

$$c_{44} U_3 \eta \cos \eta\, h = 0. \tag{3.9.5}$$

For nontrivial solutions we must have

$$\cos \eta\, h = 0, \tag{3.9.6}$$

which determines the dispersion relations of the waves as:

$$\eta h = \frac{n\pi}{2}, \quad n = 1, 3, 5, \ldots, \tag{3.9.7}$$

or

$$\eta^2 h^2 = \frac{\rho h^2 \omega^2}{c_{44}} - \xi^2 h^2 = \frac{n^2 \pi^2}{4}, \tag{3.9.8}$$

where Eq. (3.9.4) has been used. Equation (3.9.8) can be written in the dimensionless form of

$$\Omega^2 - X^2 = n^2, \tag{3.9.9}$$

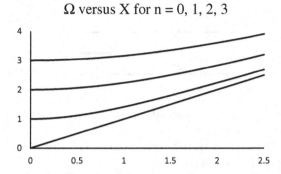

Fig. 3.5.   Dispersion curves from Eq. (3.9.9) with $n = 0, 1, 2$ and 3.

where the dimensionless frequency Ω and the dimensionless wave number X in the x_1 direction are defined by

$$\Omega = \omega \Big/ \left(\frac{\pi}{2h} \sqrt{\frac{c_{44}}{\rho}} \right), \quad X = \xi \Big/ \left(\frac{\pi}{2h} \right). \tag{3.9.10}$$

For symmetric waves we consider

$$u_3 = U_3 \cos \eta \, x_2 \cos(\xi x_1 - \omega t). \tag{3.9.11}$$

Through a similar analysis it also leads to Eq. (3.9.9) with n assuming even integers. The dispersion curves described by Eq. (3.9.9) are shown in Fig. 3.5.

3.10   Love Wave

Consider antiplane motions of an elastic plate on an elastic half-space of another material as shown in Fig. 3.6. The governing equations for the half-space and the plate are, respectively,

$$\begin{aligned}
c_{44} \nabla^2 u_3 &= \rho \ddot{u}_3, \quad x_2 > 0, \\
\hat{c}_{44} \nabla^2 u_3 &= \hat{\rho} \ddot{u}_3, \quad -h < x_2 < 0,
\end{aligned} \tag{3.10.1}$$

where $\hat{\rho}$ and $\hat{c}_{44}$ are the mass density and shear constant of the plate. We look for solutions satisfying

$$u_3 \to 0, \quad x_2 \to +\infty. \tag{3.10.2}$$

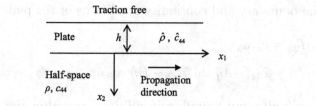

Fig. 3.6.   A plate on a half-space of a different material.

For the half-space in $x_2 > 0$, the solutions to Eq. $(3.10.1)_1$ satisfying Eq. (3.10.2) can be written as

$$u_3 = A \exp(-\eta x_2) \cos(\xi x_1 - \omega t). \qquad (3.10.3)$$

For Eq. (3.10.3) to satisfy Eq. $(3.10.1)_1$, the following must be true:

$$\eta^2 = \xi^2 - \frac{\rho \omega^2}{c_{44}} = \xi^2 \left(1 - \frac{v^2}{v_T^2} \right) > 0, \qquad (3.10.4)$$

where

$$v^2 = \frac{\omega^2}{\xi^2}, \quad v_T^2 = \frac{c_{44}}{\rho}. \qquad (3.10.5)$$

The stress component needed for the interface continuity condition is given by

$$T_{23} = c_{44} u_{3,2} = -c_{44} A \eta \exp(-\eta x_2) \cos(\xi x_1 - \omega t). \qquad (3.10.6)$$

For the plate in $-h < x_2 < 0$, let

$$u_3 = (\hat{A} \cos \hat{\eta} \, x_2 + \hat{B} \sin \hat{\eta} \, x_2) \cos(\xi x_1 - \omega t), \qquad (3.10.7)$$

where, from Eq. $(3.10.1)_2$,

$$\hat{\eta}^2 = \frac{\hat{\rho} \omega^2}{\hat{c}} - \xi^2 = \xi^2 \left(\frac{v^2}{\hat{v}_T^2} - 1 \right), \qquad (3.10.8)$$

and

$$\hat{v}_T^2 = \frac{\hat{c}}{\hat{\rho}}. \qquad (3.10.9)$$

For the boundary and continuity conditions of the plate, we need,

$$T_{23} = \hat{c}_{44} u_{3,2}$$
$$= \hat{c}_{44}(-\hat{A}\hat{\eta}\sin\hat{\eta}\,x_2 + \hat{B}\hat{\eta}\cos\hat{\eta}\,x_2)\cos(\xi x_1 - \omega t). \quad (3.10.10)$$

The continuity and boundary conditions are, after the cancellation of a common factor of $\cos(\xi x_1 - \omega t)$,

$$u_3(0^+) = A = \hat{A} = u_3(0^-),$$
$$T_{23}(0^+) = -c_{44}A\eta = \hat{c}_{44}\hat{B}\hat{\eta} = T_{23}(0^-),$$
$$T_{23}(-h) = \hat{c}_{44}(\hat{A}\hat{\eta}\sin\hat{\eta}\,h + \hat{B}\hat{\eta}\cos\hat{\eta}\,h) = 0. \quad (3.10.11)$$

Using Eqs. $(3.10.11)_1$ to eliminate A, we obtain

$$-c_{44}\hat{A}\eta - \hat{c}_{44}\hat{B}\hat{\eta} = 0,$$
$$\hat{A}\hat{\eta}\sin\hat{\eta}\,h + \hat{B}\hat{\eta}\cos\hat{\eta}\,h = 0. \quad (3.10.12)$$

For nontrivial solutions,

$$\begin{vmatrix} -c_{44}\eta & -\hat{c}_{44}\hat{\eta} \\ \hat{\eta}\sin\hat{\eta}\,h & \hat{\eta}\cos\hat{\eta}\,h \end{vmatrix} = 0, \quad (3.10.13)$$

or

$$\frac{\eta}{\xi} - \frac{\hat{c}_{44}}{c_{44}}\frac{\hat{\eta}}{\xi}\tan\hat{\eta}\,h = 0. \quad (3.10.14)$$

Substituting from Eqs. (3.10.4) and (3.10.8), we obtain

$$\sqrt{1 - \frac{v^2}{v_T^2}} - \frac{\hat{c}_{44}}{c_{44}}\sqrt{\frac{v^2}{\hat{v}_T^2} - 1}\,\tan\left[\xi h\sqrt{\frac{v^2}{\hat{v}_T^2} - 1}\right] = 0, \quad (3.10.15)$$

which determines the Love wave speed v as a function of the wave number ξ. The waves determined by Eq. (3.10.15) are dispersive. The dispersion relations for Love waves are real and multi-valued when $\hat{v}_T^2 < v^2 < v_T^2$, i.e., the material constants must be such that the shear wave speed of the plate is smaller than that of the half-space.

3.11   Small Deformations on a Finite Bias

Consider the following three states of the elastic body in Fig. 3.7. There are three coincident Cartesian coordinate systems. X_K are for the reference state. Greek coordinates and indices ξ_α are for the initial state. y_i are for the present state.

(i) In the reference state, the body is undeformed and free of loads. A generic point in this state is denoted by $\mathbf{X}$ with Cartesian coordinates X_K. The mass density is ρ^0.

(ii) The initial state is static and has finite deformations. It is also called the biasing state. Fields of the initial state are indicated by a superscript "1". In this state the body is under the action of body force $\mathbf{f}^1$, prescribed surface position $\bar{\boldsymbol{\xi}}$ and surface traction $\bar{T}_\alpha^1$. The displacement field from the reference state to the initial state is denoted by $\mathbf{w}(\mathbf{X})$. The position of the material point associated with $\mathbf{X}$ is given by $\boldsymbol{\xi} = \boldsymbol{\xi}(\mathbf{X})$ or $\xi_\alpha = \xi_\alpha(\mathbf{X})$. The strain at the initial state is E_{KL}^1. $\boldsymbol{\xi}(\mathbf{X})$ satisfies the following static equations of nonlinear elasticity:

$$J^1 = \det(\xi_{\alpha,K}),$$

$$E_{KL}^1 = (\xi_{\alpha,K}\xi_{\alpha,L} - \delta_{KL})/2 = E_{LK}^1, \qquad (3.11.1)$$

$$P_{KL}^1 = \rho^0 \left.\frac{\partial \varepsilon}{\partial E_{KL}}\right|_{\mathbf{E}^1} = P_{LK}^1, \qquad (3.11.2)$$

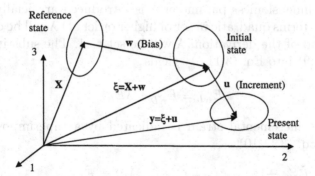

Fig. 3.7.   Reference, initial and present states of an elastic body.

$$K^1_{K\alpha} = \xi_{\alpha,L} P^1_{KL}, \tag{3.11.3}$$

$$K^1_{K\alpha,K} + \rho^0 f^1_\alpha = 0. \tag{3.11.4}$$

(iii) In the present state, the body is under the action of time-dependent loads f_i inside the body as well as $\bar{y}_i$ and $\bar{T}_i$ on the boundary surface. The present position of the material particle associated with $\mathbf{X}$ is $\mathbf{y} = \boldsymbol{\xi} + \mathbf{u}$. We consider the case when the incremental displacement $\mathbf{u}$ is infinitesimal. $\mathbf{u}$ is a function of $\boldsymbol{\xi}$ and time. Since $\boldsymbol{\xi} = \boldsymbol{\xi}(\mathbf{X})$, $\mathbf{u}$ may be viewed as a function of $\mathbf{X}$ and time. Hence $\mathbf{y}$ of the present state may also be viewed as a function of $\mathbf{X}$ and time. $\mathbf{Y}(\mathbf{X},t)$ satisfies the following dynamic equations of nonlinear elasticity:

$$J = \det(y_{i,K}),$$
$$E_{KL} = (y_{i,K} y_{i,L} - \delta_{KL})/2, \tag{3.11.5}$$

$$P_{KL} = \rho^0 \left. \frac{\partial \varepsilon}{\partial E_{KL}} \right|_{\mathbf{E}}, \tag{3.11.6}$$

$$K_{Lj} = y_{j,K} P_{KL}, \tag{3.11.7}$$

$$K_{Lj,L} + \rho^0 f_j = \rho^0 \ddot{y}_j. \tag{3.11.8}$$

Our goal is to derive linear equations governing the small and dynamic displacement $\mathbf{u}(\mathbf{X},t)$ [15,16]. To show the smallness of $\mathbf{u}$ explicitly, we write $\mathbf{y}$ as

$$y_i(\mathbf{X}, t) = \delta_{i\alpha}[\xi_\alpha(\mathbf{X}, t) + \lambda u_\alpha(\mathbf{X}, t)], \tag{3.11.9}$$

where a dimensionless parameter λ is introduced artificially. In the following, terms quadratic in (or of higher order of) λ will be dropped. At the end of the derivation, λ will be set to 1. The substitution of Eq. (3.11.9) into Eq. (3.11.5)$_2$ yields

$$E_{KL} \cong E^1_{KL} + \lambda \tilde{E}_{KL}, \tag{3.11.10}$$

where the incremental strain is indicated by a superimposed tilde "$\sim$". In Eq. (3.11.10),

$$\tilde{E}_{KL} = (\xi_{\alpha,K} u_{\alpha,L} + \xi_{\alpha,L} u_{\alpha,K})/2 = \tilde{E}_{LK}. \tag{3.11.11}$$

Further substitution of Eq. (3.11.10) into Eq. (3.11.6) gives

$$P_{KL} \cong P^1_{KL} + \lambda \tilde{P}_{KL}, \tag{3.11.12}$$

where

$$\tilde{P}_{KL} = \rho^0 \left. \frac{\partial^2 \varepsilon}{\partial E_{KL} \partial E_{MN}} \right|_{\mathbf{E}^1} \tilde{E}_{MN} = \tilde{P}_{LK}. \tag{3.11.13}$$

Then, from Eq. (3.11.7), we have

$$K_{Ki} \cong \delta_{i\alpha}(K^1_{K\alpha} + \lambda \tilde{K}_{K\alpha}), \tag{3.11.14}$$

where

$$\tilde{K}_{K\alpha} = u_{\alpha,L} P^1_{KL} + \xi_{\alpha,L} \tilde{P}_{KL}. \tag{3.11.15}$$

With Eq. (3.11.13), we can write Eq. (3.11.15) as

$$\tilde{K}_{L\gamma} = G_{L\gamma M\alpha} u_{\alpha,M}, \tag{3.11.16}$$

where

$$G_{K\alpha L\gamma} = \xi_{\alpha,M} \, \rho^0 \left. \frac{\partial^2 \varepsilon}{\partial E_{KM} \partial E_{LN}} \right|_{\mathbf{E}^1} \xi_{\gamma,N} + P^1_{KL} \delta_{\alpha\gamma}$$

$$= G_{L\gamma K\alpha}. \tag{3.11.17}$$

Equation (3.11.16) shows that the incremental stress tensor depends on the incremental displacement gradients linearly. $G_{K\alpha L\gamma}$ are referred to as the effective or apparent elastic constants of the material under a bias. They depend on the initial deformation $\xi_\alpha(\mathbf{X})$.

Since the fields in the present state satisfy Eq. (3.11.8) and the biasing fields satisfy Eq. (3.11.4), we have

$$\tilde{K}_{K\alpha,K} + \rho^0 \tilde{f}_\alpha = \rho^0 \ddot{u}_\alpha, \qquad (3.11.18)$$

where $\tilde{f}_\alpha$ is given by

$$f_i = \delta_{i\alpha}(f_\alpha^1 + \lambda \tilde{f}_\alpha). \qquad (3.11.19)$$

In some applications, the biasing deformations are also infinitesimal and are governed by the linear theory of elasticity:

$$T_{KL,K}^1 + \rho^0 f_L^1 = 0, \qquad (3.11.20)$$

$$T_{KL}^1 = c_{KLMN} E_{MN}^1 = T_{LK}^1, \qquad (3.11.21)$$

$$E_{KL}^1 = \frac{1}{2}(w_{K,L} + w_{L,K}) = E_{LK}^1. \qquad (3.11.22)$$

Since the biasing deformations are infinitesimal, we consider their first-order effects on the incremental fields only. In this case the following internal energy density is sufficient:

$$\rho^0 \varepsilon = \frac{1}{2} c_{ABCD} E_{AB} E_{CD} + \frac{1}{6} c_{ABCDEF} E_{AB} E_{CD} E_{EF}, \qquad (3.11.23)$$

where we have simplified the notation of the third-order material constants and denoted

$$c_{ABCDEF} = \underset{3}{c}_{ABCDEF}. \qquad (3.11.24)$$

Then, neglecting the second-order terms of the gradients of the small **w**, we can write the effective elastic constants as:

$$G_{K\alpha L\gamma} = c_{K\alpha L\gamma} + \hat{c}_{K\alpha L\gamma}, \qquad (3.11.25)$$

where

$$\hat{c}_{K\alpha L\gamma} = T_{KL}^1 \delta_{\alpha\gamma} + c_{K\alpha LN} w_{\gamma,N} + c_{KNL\gamma} w_{\alpha,N}$$

$$+ c_{K\alpha L\gamma AB} E_{AB}^1 = \hat{c}_{L\gamma K\alpha}. \qquad (3.11.26)$$

Equation (3.11.26) shows explicitly that $G_{K\alpha L\gamma}$ depends on the small and initial deformations linearly. When the initial deformation is nonuniform, the equations for the incremental deformations are with variable coefficients. The effective elastic constants $G_{K\alpha L\gamma}$ in general have lower symmetry than the fundamental linear elastic constants $c_{K\alpha L\gamma}$. This is called induced anisotropy or symmetry breaking due to biasing deformations. The third-order elastic constants in Eq. (3.11.26) are necessary for a complete description of the first-order effects of the biasing deformations.

3.12 Thermal and Dissipative Effects

Thermal and dissipative effects often appear together and are treated in this section [17]. For this purpose the energy equation in Eq. (3.1.4) in integral form needs to be extended to include thermal effects:

$$\frac{d}{dt}\int_v \rho\left(\frac{1}{2}\mathbf{v}\cdot\mathbf{v}+\varepsilon\right)dv$$

$$= \int_v (\rho\mathbf{f}\cdot\mathbf{v}+\rho r)dv + \int_s (\mathbf{t}\cdot\mathbf{v}-\mathbf{n}\cdot\mathbf{q})ds, \qquad (3.12.1)$$

where $\mathbf{q}$ is the heat flux vector and r is the body heat source per unit mass. The second law of thermodynamics needs to be added as follows:

$$\frac{d}{dt}\int_v \rho\eta\, dv \geq \int_v \frac{\rho r}{\theta}dv - \int_s \frac{\mathbf{q}\cdot\mathbf{n}}{\theta}ds, \qquad (3.12.2)$$

where η is the entropy per unit mass and θ is the absolute temperature. Equations (3.12.1) and (3.12.2) can be converted to differential forms using the divergence theorem in Appendix C. The results are

$$\rho\dot{\varepsilon} = \tau_{ij}v_{j,i} + \rho r - q_{i,i}, \qquad (3.12.3)$$

$$\rho\dot{\eta} \geq \frac{\rho r}{\theta} - \left(\frac{q_i}{\theta}\right)_{,i}. \qquad (3.12.4)$$

Eliminating r from Eqs. (3.12.3) and (3.12.4), we obtain the Clausius–Duhem inequality as

$$\rho(\theta\dot{\eta} - \dot{\varepsilon}) + \tau_{ij}v_{j,i} - \frac{q_i\theta_{,i}}{\theta} \geq 0. \qquad (3.12.5)$$

A free energy F can be introduced through the following Legendre transform of the internal energy ε:

$$F = \varepsilon - \theta\eta,$$
$$\dot{F} = \dot{\varepsilon} - \dot{\theta}\eta - \theta\dot{\eta}. \tag{3.12.6}$$

Then the energy equation in Eq. (3.12.3) and the Clausius–Duhem inequality in Eq. (3.12.5) become

$$\rho(\dot{F} + \eta\dot{\theta} + \dot{\eta}\theta) = \tau_{ij}v_{j,i} + \rho r - q_{i,i}, \tag{3.12.7}$$

and

$$-\rho(\dot{F} + \eta\dot{\theta}) + \tau_{ij}v_{j,i} - \frac{q_i\theta_{,i}}{\theta} \geq 0. \tag{3.12.8}$$

We introduce the following material heat flux and material temperature gradient:

$$Q_K = JX_{K,k}q_k, \quad \Theta_K = \theta_{,K} = \theta_{,k}y_{k,K}. \tag{3.12.9}$$

Then Eqs. (3.12.7) and (3.12.8) can be written in material forms as

$$\rho^0(\dot{F} + \eta\dot{\theta} + \dot{\eta}\theta) = P_{KL}\dot{E}_{KL} + \rho^0 r - Q_{K,K}, \tag{3.12.10}$$

$$-\rho^0(\dot{F} + \eta\dot{\theta}) + P_{KL}\dot{E}_{KL} - \frac{Q_K\Theta_K}{\theta} \geq 0. \tag{3.12.11}$$

For constitutive relations we start with the following expressions:

$$F = F(E_{KL}; \theta; \Theta_K),$$
$$P_{KL} = P_{KL}(E_{KL}; \theta; \Theta_K; \dot{E}_{KL}), \tag{3.12.12}$$
$$Q_K = Q_K(E_{KL}; \theta; \Theta_K; \dot{E}_{KL}).$$

The substitution of Eq. (3.12.12) into the Clausius–Duhem inequality in Eq. (3.12.11) yields

$$-\rho^0 \frac{\partial F}{\partial \Theta_K}\dot{\Theta}_K - \rho^0\left(\eta + \frac{\partial F}{\partial \theta}\right)\dot{\theta}$$

$$+ \left(P_{KL} - \rho^0 \frac{\partial F}{\partial E_{KL}}\right)\dot{E}_{KL} - \frac{1}{\theta}Q_K\Theta_K \geq 0. \tag{3.12.13}$$

Since Eq. (3.12.13) is linear in $\dot{\Theta}_K$ and $\dot{\theta}$, for the inequality to hold F cannot depend on Θ_K, and η has to be related to F by

$$\eta = -\frac{\partial F}{\partial \theta}. \tag{3.12.14}$$

In addition, we break P_{KL} into reversible and dissipative parts as follows:

$$P_{KL} = P_{KL}^R + P_{KL}^D, \tag{3.12.15}$$

where

$$P_{KL}^R = \rho^0 \frac{\partial F}{\partial E_{KL}}, \quad F = F(E_{KL}; \theta), \tag{3.12.16}$$

$$P_{KL}^D = P_{KL}^D(E_{KL}; \theta; \Theta_K; \dot{E}_{KL}). \tag{3.12.17}$$

Then what is left from the Clausius–Duhem inequality in Eq. (3.12.13) is

$$P_{KL}^D \dot{E}_{KL} - \frac{1}{\theta} Q_K \Theta_K \geq 0. \tag{3.12.18}$$

From the energy equation in Eq. (3.12.10) we obtain the heat equation or dissipation equation as

$$\rho^0 \theta \dot{\eta} = P_{KL}^D \dot{E}_{KL} + \rho^0 r - Q_{K,K}, \tag{3.12.19}$$

where Eq. (3.12.16) has been used.

In summary, the nonlinear equations of thermoviscoelasticity are

$$\rho^0 = \rho J,$$
$$K_{Lk,L} + \rho^0 f_k = \rho^0 \ddot{y}_k, \tag{3.12.20}$$
$$\rho^0 \theta \dot{\eta} = P_{KL}^D \dot{E}_{KL} + \rho_0 r - Q_{K,K},$$

with constitutive relations given by Eqs. (3.12.14)–(3.12.17) and Eq. (3.12.12)$_3$ which are restricted by Eq. (3.12.18). The equation for the conservation of mass in Eq. (3.12.20)$_1$ can be used to determine

ρ separately from the other equations. Equations $(3.12.20)_{2,3}$ can be written as four equations for $y_i(\mathbf{X},t)$ and $\theta(\mathbf{X},t)$. On the boundary surface S, the thermal boundary condition may be either prescribed temperature or heat flux

$$N_L Q_L = \bar{Q}. \tag{3.12.21}$$

For jump conditions at a nonmaterial discontinuity surface in general, we need to go back to Eqs. (3.12.1) and (3.12.2). The energy equation in Eq. (3.12.1) is different from Eq. (3.1.4) because of the thermal fields:

$$\frac{d}{dt}\int_v \rho\left(\frac{1}{2}\mathbf{v}\cdot\mathbf{v}+\varepsilon\right)dv = \int_v (\rho\mathbf{f}\cdot\mathbf{v}+\rho r)dv$$

$$+ \int_s (\mathbf{t}\cdot\mathbf{v}-\mathbf{n}\cdot\mathbf{q})ds. \tag{3.12.22}$$

Comparing Eq. (3.12.22) with the general balance law in Eq. (3.3.1):

$$\frac{d}{dt}\int_v \rho\Psi dv = \int_v \rho\Gamma dv + \int_s \mathbf{n}\cdot\mathbf{\Phi}ds, \tag{3.12.23}$$

we identify

$$\Psi = \frac{1}{2}\mathbf{v}\cdot\mathbf{v}+\varepsilon,$$

$$\Gamma = \mathbf{f}\cdot\mathbf{v}+r, \quad \mathbf{\Phi} = \boldsymbol{\tau}\cdot\mathbf{v}-\mathbf{q}. \tag{3.12.24}$$

From the general jump condition in Eq. (3.3.10), we have

$$[\rho\Psi\Theta + \mathbf{n}\cdot\mathbf{\Phi}] = 0, \tag{3.12.25}$$

where

$$\Theta = N - \mathbf{v}\cdot\mathbf{n}. \tag{3.12.26}$$

The substitution of Eq. (3.12.24) into Eq. (3.12.25) gives the jump condition associated with the energy equation:

$$[\rho(\mathbf{v}\cdot\mathbf{v}/2+\varepsilon)\Theta + \mathbf{n}\cdot(\boldsymbol{\tau}\cdot\mathbf{v}-\mathbf{q})] = 0. \tag{3.12.27}$$

In the case of a material surface with $\Theta = 0$, Eq. (3.12.27) reduces to

$$[\mathbf{n}\cdot\mathbf{q}] = 0. \tag{3.12.28}$$

For the second law of thermodynamics in Eq. (3.12.2):

$$\frac{d}{dt}\int_v \rho\eta dv \geq \int_v \frac{\rho r}{\theta}dv - \int_s \frac{\mathbf{q}\cdot\mathbf{n}}{\theta}ds, \qquad (3.12.29)$$

we need the transport theorem and divergence theorem in Eqs. (2.7.12) and (2.7.9) which, with $\phi = \rho\eta$ and $\mathbf{\Phi} = \mathbf{q}/\theta$, take the following form:

$$\frac{d}{dt}\int_v \rho\eta dv = \int_{v-\Sigma}\left\{\frac{\partial(\rho\eta)}{\partial t} + (\rho\eta v_k)_{,k}\right\}dv$$

$$- \int_\Sigma [\rho\eta\Theta]ds, \qquad (3.12.30)$$

$$\int_s \mathbf{n}\cdot\mathbf{q}/\theta ds = \int_{v-\Sigma}\nabla\cdot(\mathbf{q}/\theta)dv + \int_\Sigma \mathbf{n}\cdot[\mathbf{q}/\theta]ds. \quad (3.12.31)$$

Substituting Eqs. (3.12.30) and (3.12.31) into Eq. (3.12.29), we obtain

$$\int_{v-\Sigma}\left\{\frac{\partial(\rho\eta)}{\partial t} + (\rho\eta v_k)_{,k}\right\}dv - \int_\Sigma [\rho\eta\Theta]ds$$

$$\geq \int_v \frac{\rho r}{\theta}dv - \int_{v-\Sigma}\nabla\cdot(\mathbf{q}/\theta)dv - \int_\Sigma \mathbf{n}\cdot[\mathbf{q}/\theta]ds,$$

$$(3.12.32)$$

which can be written as

$$\int_{v-\Sigma}\left\{\frac{\partial(\rho\eta)}{\partial t} + (\rho\eta v_k)_{,k} - \frac{\rho r}{\theta} + \left(\frac{q_k}{\theta}\right)_{,k}\right\}dv$$

$$+ \int_\Sigma [\mathbf{n}\cdot\mathbf{q}/\theta - \rho\eta\Theta]ds \geq 0, \qquad (3.12.33)$$

or

$$\int_{v-\Sigma}\left\{\rho\frac{d\eta}{dt} - \frac{\rho r}{\theta} + \left(\frac{q_k}{\theta}\right)_{,k}\right\}dv$$

$$+ \int_\Sigma [\mathbf{n}\cdot\mathbf{q}/\theta - \rho\eta\Theta]ds \geq 0. \qquad (3.12.34)$$

To ensure that Eq. (3.12.34) is satisfied,

$$\rho\frac{d\eta}{dt} - \frac{\rho r}{\theta} + \left(\frac{q_k}{\theta}\right)_{,k} \geq 0 \quad \text{in} \quad v - \Sigma, \qquad (3.12.35)$$

$$[\mathbf{n} \cdot \mathbf{q}/\theta - \rho\eta\Theta] \geq 0 \quad \text{on} \quad \Sigma. \qquad (3.12.36)$$

Equation (3.12.35) is Eq. (3.12.4). Equation (3.12.36) is the jump condition associated with the second law of thermodynamics. In the case of a material surface for which $\Theta = 0$, assuming θ is continuous across the material surface, Eq. (3.12.36) reduces to

$$[\mathbf{n} \cdot \mathbf{q}] \geq 0 \quad \text{on} \quad \Sigma, \qquad (3.12.37)$$

which is ensured by Eq. (3.12.28).

3.13 Cylindrical Coordinates

To analyze circular cylindrical structures it is convenient to use cylindrical coordinates (r, θ, z) defined by

$$x_1 = r\cos\theta, \quad x_2 = r\sin\theta, \quad x_3 = z. \qquad (3.13.1)$$

In cylindrical coordinates, we have the following strain–displacement relations:

$$S_{rr} = u_{r,r}, \quad S_{\theta\theta} = \frac{1}{r}u_{\theta,\theta} + \frac{u_r}{r}, \quad S_{zz} = u_{z,z},$$

$$2S_{r\theta} = u_{\theta,r} + \frac{1}{r}u_{r,\theta} - \frac{u_\theta}{r}, \quad 2S_{\theta z} = \frac{1}{r}u_{z,\theta} + u_{\theta,z}, \qquad (3.13.2)$$

$$2S_{zr} = u_{r,z} + u_{z,r}.$$

The equations of motion take the following form:

$$\frac{\partial T_{rr}}{\partial r} + \frac{1}{r}\frac{\partial T_{\theta r}}{\partial\theta} + \frac{\partial T_{zr}}{\partial z} + \frac{T_{rr} - T_{\theta\theta}}{r} + f_r = \rho\ddot{u}_r,$$

$$\frac{\partial T_{r\theta}}{\partial r} + \frac{1}{r}\frac{\partial T_{\theta\theta}}{\partial\theta} + \frac{\partial T_{z\theta}}{\partial z} + \frac{2}{r}T_{r\theta} + f_\theta = \rho\ddot{u}_\theta, \qquad (3.13.3)$$

$$\frac{\partial T_{rz}}{\partial r} + \frac{1}{r}\frac{\partial T_{\theta z}}{\partial\theta} + \frac{\partial T_{zz}}{\partial z} + \frac{1}{r}T_{rz} + f_z = \rho\ddot{u}_z.$$

The gradient of a scalar field ψ is given by

$$\nabla \psi = \frac{\partial \psi}{\partial r} \mathbf{e}_r + \frac{1}{r} \frac{\partial \psi}{\partial \theta} \mathbf{e}_\theta + \frac{\partial \psi}{\partial z} \mathbf{e}_z. \tag{3.13.4}$$

The divergence of a vector field $\mathbf{B}$ is

$$\nabla \cdot \mathbf{B} = \frac{1}{r} (r B_r)_{,r} + \frac{1}{r} B_{\theta,\theta} + B_{z,z}. \tag{3.13.5}$$

The Laplace operator (Laplacian) on a scalar field ψ takes the following form:

$$\nabla^2 \psi = \frac{1}{r} \frac{\partial}{\partial r} \left(r \frac{\partial \psi}{\partial r} \right) + \frac{1}{r^2} \frac{\partial^2 \psi}{\partial \theta^2} + \frac{\partial^2 \psi}{\partial z^2}. \tag{3.13.6}$$

Chapter 4

Electromagnetoelastic
Interaction Model

When a deformable, polarizable and magnetizable material is subjected to an electromagnetic field, a differential element of the material experiences an electromagnetic body force and a couple. When such a material polarizes and magnetizes, the electromagnetic field does work on the material. Fundamental to the development of the macroscopic equations of electromagnetoelasticity is the derivation of the electromagnetic body force, couple and power due to the electromagnetic field. This can be done using the charged particle model in [18,19]. In this book, we use the two-continuum model constructed by Tiersten [17,20]. The model can be extended to describe other electromagnetoelastic solids such as semiconductors and ferromagnets [21,22].

4.1 Two-Continuum Model

Following [17,20], an electromagnetoelastic body is modeled by a lattice continuum and a bound charge continuum. In the reference state without any deformation and fields, the two continua coincide with each other. The reference position of a typical material particle of the lattice continuum is $\mathbf{X}$ and its current position is $\mathbf{y} = \mathbf{y}(\mathbf{X}, t)$. The bound charge continuum can displace a little from the lattice continuum by an infinitesimal displacement $\boldsymbol{\eta}$. It is assumed that $\boldsymbol{\eta}(\mathbf{y}, t)$ preserves the volume of the bound charge continuum, i.e.,

$$\eta_{k,k} = \frac{\partial \eta_k}{\partial y_k} = \nabla \cdot \boldsymbol{\eta} = 0, \tag{4.1.1}$$

where

$$\nabla = \mathbf{e}_k \frac{\partial}{\partial y_k}. \tag{4.1.2}$$

Equation (4.1.1) is needed to obtain the proper electric charge equation (Gauss' law) [21], which will be shown later.

We consider insulators without free charges and their currents. In the reference state, the lattice continuum has a mass density $\rho^0(\mathbf{X})$ and a positive charge density $_0\mu^l(\mathbf{X})$. The bound charge continuum is massless and has a negative charge density $_0\mu^b(\mathbf{X})$. At the reference state, the body is assumed to be electrically neutral with

$$_0\mu^l(\mathbf{X}) + _0\mu^b(\mathbf{X}) = 0. \tag{4.1.3}$$

In the current state, the mass density of the lattice continuum is ρ and its charge density is μ^l. The current charge density of the bound charge continuum is μ^b. As a consequence of Eqs. (4.1.1) and (4.1.3), we have the following charge neutrality condition at the current state:

$$\mu^l(\mathbf{y}, t) + \mu^b(\mathbf{y} + \boldsymbol{\eta}, t) = 0. \tag{4.1.4}$$

The bound charge continuum also carries a magnetic moment density $\mathbf{M}'$ about the point at $\mathbf{y}$ per unit volume in the instantaneous local rest frame R_C at $\mathbf{y}$:

$$\mathbf{M}' = \frac{1}{2} \int_{c'} \mathbf{r} \times I' \mathbf{dl}. \tag{4.1.5}$$

The $\mathbf{M}'$ in R_C at $\mathbf{y}$ is related to the $\mathbf{M}$ in the fixed reference frame R_G by

$$\mathbf{M}' = \mathbf{M} + \mathbf{v} \times \mathbf{P}. \tag{4.1.6}$$

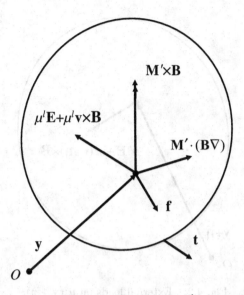

Fig. 4.1.   External loads on the lattice, $\mu^l(\mathbf{y})$ and $\mathbf{M}'(\mathbf{y})$.

At $\mathbf{y}$, the lattice continuum is under the usual mechanical body force $\mathbf{f}$ per unit mass as shown in Fig. 4.1. The lattice continuum is also under the surface traction $\mathbf{t}$ per unit area on its boundary surface. μ^l of the lattice experiences the following electromagnetic body force at $\mathbf{y}$ (Lorentz force):

$$\mu^l \mathbf{E} + \mu^l \mathbf{v} \times \mathbf{B}. \qquad (4.1.7)$$

In addition, there is a magnetic body force on the $\mathbf{M}'$ at $\mathbf{y}$ per unit volume:

$$\int_{c'} I' \mathbf{dl} \times \mathbf{B} = \mathbf{M}' \cdot (\mathbf{B}\nabla). \qquad (4.1.8)$$

The $\mathbf{M}'$ at $\mathbf{y}$ also experiences a magnetic body couple per unit volume:

$$\int_{c'} \mathbf{r} \times (I' \mathbf{dl}) \times \mathbf{B} = \mathbf{M}' \times \mathbf{B}. \qquad (4.1.9)$$

Similarly, at $\mathbf{y} + \boldsymbol{\eta}$, there is an electromagnetic body force on μ^b as shown in Fig. 4.2:

$$\mu^b \mathbf{E} + \mu^b (\mathbf{v} + \dot{\boldsymbol{\eta}}) \times \mathbf{B}. \qquad (4.1.10)$$

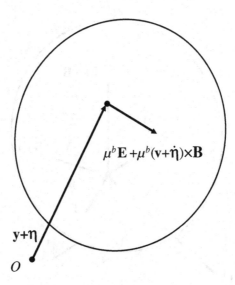

Fig. 4.2.   External loads on $\mu^b(\mathbf{y} + \boldsymbol{\eta})$.

The interactions between the two continua [17,20] are not shown in Figs. 4.1 and 4.2 because they are internal and will cancel with each other when the two continua are treated together later.

In terms of $\boldsymbol{\eta}$, the polarization per unit volume is defined by [17]

$$\mathbf{P} = \mu^l(\mathbf{y})(-\boldsymbol{\eta}) = \mu^b(\mathbf{y}+\boldsymbol{\eta})\boldsymbol{\eta} \cong \mu^b(\mathbf{y})\boldsymbol{\eta}. \qquad (4.1.11)$$

Similar to the equation for the conservation of mass, i.e.,

$$\frac{d\rho}{dt} + \rho v_{k,k} = 0, \qquad (4.1.12)$$

we also have the following equations from the conservation of charge for the lattice and bound charge continua, respectively:

$$\frac{d\mu^l}{dt} + \mu^l v_{k,k} = 0, \qquad (4.1.13)$$

$$\frac{d\mu^b}{dt} + \mu^b(v_k + \dot{\eta}_k)_{,k} = \frac{d\mu^b}{dt} + \mu^b v_{k,k} = 0. \qquad (4.1.14)$$

From Eqs. (4.1.12) to (4.1.14) we obtain the following relationship which will be useful later:

$$\frac{\dot{\mu}^l}{\mu^l} = \frac{\dot{\mu}^b}{\mu^b} = \frac{\dot{\rho}}{\rho} = -v_{k,k}. \tag{4.1.15}$$

For later convenience, we introduce the polarization per unit mass by

$$\boldsymbol{\pi} = \frac{\mathbf{P}}{\rho}. \tag{4.1.16}$$

We have

$$\mu^b \dot{\boldsymbol{\eta}} = \frac{d}{dt}(\mu^b \boldsymbol{\eta}) - \dot{\mu}^b \boldsymbol{\eta}$$

$$= \dot{\mathbf{P}} - \frac{\mathbf{P}}{\mu^b} \dot{\mu}^b = \dot{\mathbf{P}} - \mathbf{P}\frac{\dot{\rho}}{\rho}$$

$$= \rho \dot{\boldsymbol{\pi}} + \dot{\rho}\boldsymbol{\pi} - \dot{\rho}\boldsymbol{\pi} = \rho \dot{\boldsymbol{\pi}}. \tag{4.1.17}$$

Gauss' law for the electric field can be written as

$$\int_s \varepsilon_0 \mathbf{n} \cdot \mathbf{E} ds = \int_v [\mu^l(\mathbf{y}) + \mu^b(\mathbf{y})] dv. \tag{4.1.18}$$

The differential form of Eq. (4.1.18) is [21]

$$\varepsilon_0 E_{k,k} = \mu^l(\mathbf{y}) + \mu^b(\mathbf{y}) = -\mu^b(\mathbf{y} + \boldsymbol{\eta}) + \mu^b(\mathbf{y})$$

$$= -\mu^b(\mathbf{y}) - \mu^b_{,i}(\mathbf{y})\eta_i + \mu^b(\mathbf{y})$$

$$= -[\mu^b(\mathbf{y})\eta_i]_{,i} + \mu^b(\mathbf{y})\eta_{i,i} = -P_{i,i}. \tag{4.1.19}$$

Equation (4.1.19) can be further written as

$$(\varepsilon_0 E_i + P_i)_{,i} = D_{i,i} = 0, \tag{4.1.20}$$

where we have introduced the electric displacement vector $\mathbf{D}$ through

$$D_i = \varepsilon_0 E_i + P_i. \tag{4.1.21}$$

According to Eq. (1.10.12), the electromagnetic power on $\mathbf{M}'$ is

$$w'^M = -\mathbf{M}' \cdot \frac{\partial \mathbf{B}}{\partial t}. \tag{4.1.22}$$

4.2 Electromagnetic Body Force

The electromagnetic body force, couple and power can be obtained from the two-continuum model in the previous section. We begin with the electromagnetic body force per unit volume. Consider a unit volume of the lattice continuum at $\mathbf{y}$. The associated bound charge continuum at $\mathbf{y} + \boldsymbol{\eta}$ also has a unit volume because of Eq. (4.1.1). From Figs. 4.1 and 4.2, the electromagnetic body force on the lattice continuum per unit volume and the associated bound charge continuum together is

$$
\begin{aligned}
\mathbf{F}^{EM} &= \mu^l(\mathbf{y})\mathbf{E}(\mathbf{y}) + \mu^l(\mathbf{y})\mathbf{v}(\mathbf{y}) \times \mathbf{B}(\mathbf{y}) \\
&\quad + \mu^b(\mathbf{y} + \boldsymbol{\eta})\mathbf{E}(\mathbf{y} + \boldsymbol{\eta}) \\
&\quad + \mu^b(\mathbf{y} + \boldsymbol{\eta})(\mathbf{v} + \dot{\boldsymbol{\eta}}) \times \mathbf{B}(\mathbf{y} + \boldsymbol{\eta}) + \mathbf{M}' \cdot (\mathbf{B}\nabla).
\end{aligned} \tag{4.2.1}
$$

Using

$$
\begin{aligned}
E_j(\mathbf{y} + \boldsymbol{\eta}) &\cong E_j(\mathbf{y}) + \eta_i E_{j,i}(\mathbf{y}), \\
B_j(\mathbf{y} + \boldsymbol{\eta}) &\cong B_j(\mathbf{y}) + \eta_i B_{j,i}(\mathbf{y}),
\end{aligned} \tag{4.2.2}
$$

we can write $\mathbf{F}^{EM}$ as

$$
\begin{aligned}
\mathbf{F}^{EM} &= [\mu^l(\mathbf{y}) + \mu^b(\mathbf{y} + \boldsymbol{\eta})]\mathbf{E}(\mathbf{y}) \\
&\quad + [\mu^l(\mathbf{y}) + \mu^b(\mathbf{y} + \boldsymbol{\eta})]\mathbf{v}(\mathbf{y}) \times \mathbf{B}(\mathbf{y}) \\
&\quad + \mathbf{P} \cdot \nabla \mathbf{E} + \mathbf{M}' \cdot (\mathbf{B}\nabla) \\
&\quad + \mathbf{v} \times (\mathbf{P} \cdot \nabla \mathbf{B}) + \rho \dot{\boldsymbol{\pi}} \times \mathbf{B} \\
&= \mathbf{P} \cdot \nabla \mathbf{E} + \mathbf{M}' \cdot (\mathbf{B}\nabla) \\
&\quad + \mathbf{v} \times (\mathbf{P} \cdot \nabla \mathbf{B}) + \rho \dot{\boldsymbol{\pi}} \times \mathbf{B},
\end{aligned} \tag{4.2.3}
$$

where a term of the product of the small $\boldsymbol{\eta}$, its material time derivative and the gradient of $\mathbf{B}$ has been omitted as an approximation [20]. In component form, Eq. (4.2.3) becomes

$$
\begin{aligned}
F_j^{EM} &= P_i E_{j,i} + M_i' B_{i,j} \\
&\quad + \varepsilon_{jkl} v_k P_i B_{l,i} + \varepsilon_{jkl} \rho \dot{\pi}_k B_l.
\end{aligned} \tag{4.2.4}
$$

$\mathbf{F}^{EM}$ can also be written as

$$
\mathbf{F}^{EM} = \mathbf{P} \cdot \nabla \mathbf{E}' + \mathbf{M}' \cdot (\mathbf{B}\nabla) + \mathbf{P}^* \times \mathbf{B}. \tag{4.2.5}
$$

4.3 Electromagnetic Body Couple

In this section, we calculate the moment of the electromagnetic forces and couple on the lattice and bound charge continua about the origin of $\mathbf{y}$. From Figs. 4.1 and 4.2, the moment on a unit lattice continuum at $\mathbf{y}$ and the associated bound charge continuum at $\mathbf{y} + \boldsymbol{\eta}$ together is given by

$$\begin{aligned}
\mathbf{y} \times [\mu^l(\mathbf{y})\mathbf{E}(\mathbf{y}) &+ \mu^l(\mathbf{y})\mathbf{v}(\mathbf{y}) \times \mathbf{B}(\mathbf{y}) + \mathbf{M}' \cdot (\mathbf{B}\nabla)] \\
&+ (\mathbf{y} + \boldsymbol{\eta}) \times [\mu^b(\mathbf{y} + \boldsymbol{\eta})\mathbf{E}(\mathbf{y} + \boldsymbol{\eta}) \\
&+ \mu^b(\mathbf{y} + \boldsymbol{\eta})(\mathbf{v} + \dot{\boldsymbol{\eta}}) \times \mathbf{B}(\mathbf{y} + \boldsymbol{\eta})] + \mathbf{M}' \times \mathbf{B} \\
&= \mathbf{y} \times \mathbf{F}^{EM} + \mathbf{C}^{EM},
\end{aligned} \tag{4.3.1}$$

where we have introduced the electromagnetic body couple $\mathbf{C}^{EM}$ per unit volume by

$$\mathbf{C}^{EM} = \mathbf{P} \times \mathbf{E}' + \mathbf{M}' \times \mathbf{B}, \tag{4.3.2}$$

or

$$C_k^{EM} = \varepsilon_{kij} P_i E_j' + \varepsilon_{kij} M_i' B_j. \tag{4.3.3}$$

In Eq. (4.3.3),

$$\begin{aligned}
\mathbf{E}' &= \mathbf{E} + \mathbf{v} \times \mathbf{B}, \\
\mathbf{M}' &\cong \mathbf{M} + \mathbf{v} \times \mathbf{P}.
\end{aligned} \tag{4.3.4}$$

$\mathbf{E}'$ and $\mathbf{M}'$ are the fields in the rest frame $\mathrm{R_C}$. In the derivation of Eq. (4.3.1), a few terms involving the products of the small $\boldsymbol{\eta}$ with itself, and/or its material time derivative, and/or the gradients of $\mathbf{E}$ or $\mathbf{B}$ have been neglected as an approximation [20]. The electromagnetic body couple can also be written as

$$\begin{aligned}
\mathbf{C}^{EM} &= \mathbf{P} \times \mathbf{E}' + \mathbf{M}' \times \mathbf{B} \\
&= \mathbf{P} \times (\mathbf{E} + \mathbf{v} \times \mathbf{B}) + (\mathbf{M} + \mathbf{v} \times \mathbf{P}) \times \mathbf{B} \\
&= \mathbf{P} \times \mathbf{E} + \mathbf{M} \times \mathbf{B} + \mathbf{v} \times (\mathbf{P} \times \mathbf{B}).
\end{aligned} \tag{4.3.5}$$

4.4   Electromagnetic Body Power

The electromagnetic power per unit volume on the lattice continuum and the associated bound charge continuum together can be
calculated from

$$
\begin{aligned}
W^{EM} &= [\mu^l(\mathbf{y})\mathbf{E}(\mathbf{y}) + \mu^l(\mathbf{y})\mathbf{v}(\mathbf{y}) \times \mathbf{B}(\mathbf{y})] \cdot \mathbf{v} \\
&\quad + [\mu^b(\mathbf{y}+\boldsymbol{\eta})\mathbf{E}(\mathbf{y}+\boldsymbol{\eta}) + \mu^b(\mathbf{y}+\boldsymbol{\eta})(\mathbf{v}+\dot{\boldsymbol{\eta}}) \\
&\quad \times \mathbf{B}(\mathbf{y}+\boldsymbol{\eta})] \cdot (\mathbf{v}+\dot{\boldsymbol{\eta}}) - \mathbf{M}' \cdot \frac{\partial \mathbf{B}}{\partial t} \\
&= [\mu^l(\mathbf{y})\mathbf{E}(\mathbf{y})] \cdot \mathbf{v} + [\mu^b(\mathbf{y}+\boldsymbol{\eta})\mathbf{E}(\mathbf{y}+\boldsymbol{\eta})] \cdot (\mathbf{v}+\dot{\boldsymbol{\eta}}) \\
&\quad - \mathbf{M}' \cdot \frac{\partial \mathbf{B}}{\partial t},
\end{aligned}
\tag{4.4.1}
$$

where Eq. (4.1.22) has been used for the last term. With some algebra, Eq. (4.4.1) can be written as

$$
\begin{aligned}
W^{EM} &= [\mu^l(\mathbf{y}) + \mu^b(\mathbf{y}+\boldsymbol{\eta})]\mathbf{E}(\mathbf{y}) \cdot \mathbf{v} \\
&\quad + (\mathbf{P} \cdot \nabla \mathbf{E}) \cdot \mathbf{v} + \rho \mathbf{E} \cdot \dot{\boldsymbol{\pi}} - \mathbf{M}' \cdot \frac{\partial \mathbf{B}}{\partial t} \\
&= (\mathbf{P} \cdot \nabla \mathbf{E}) \cdot \mathbf{v} + \rho \mathbf{E} \cdot \dot{\boldsymbol{\pi}} - \mathbf{M}' \cdot \frac{\partial \mathbf{B}}{\partial t} \\
&= P_i E_{j,i} v_j + \rho E_j \dot{\pi}_j - M'_j \frac{\partial B_i}{\partial t},
\end{aligned}
\tag{4.4.2}
$$

which can also be written as

$$
\begin{aligned}
W^{EM} &= \mathbf{F}^{EM} \cdot \mathbf{v} + \rho \mathbf{E}' \cdot \dot{\boldsymbol{\pi}} - \mathbf{M}' \cdot \dot{\mathbf{B}} \\
&= F_j^{EM} v_j + \rho E'_j \dot{\pi}_j - M'_j \dot{B}_j.
\end{aligned}
\tag{4.4.3}
$$

Another useful expression of W^{EM} is

$$
\begin{aligned}
W^{EM} &= E_i \frac{\partial P_i}{\partial t} - M_i \frac{\partial B_i}{\partial t} + (v_k P_i E_i)_{,k} \\
&= \mathbf{E} \cdot \frac{\partial \mathbf{P}}{\partial t} - \mathbf{M} \cdot \frac{\partial \mathbf{B}}{\partial t} + \nabla \cdot [\mathbf{v}(\mathbf{P} \cdot \mathbf{E})].
\end{aligned}
\tag{4.4.4}
$$

4.5   Energy and Momentum

From the electromagnetic body force

$$F_j^{EM} = P_i E_{j,i} + M_i' B_{i,j}$$
$$+ \varepsilon_{jkl} v_k P_i B_{l,i} + \varepsilon_{jkl} \rho \dot{\pi}_k B_l, \tag{4.5.1}$$

where

$$\mathbf{E}' = \mathbf{E} + \mathbf{v} \times \mathbf{B},$$
$$\mathbf{M}' \cong \mathbf{M} + \mathbf{v} \times \mathbf{P}, \tag{4.5.2}$$
$$\boldsymbol{\pi} = \mathbf{P}/\rho,$$

through some algebra, it can be shown that [20]

$$F_j^{EM} = T_{ij,i}^{EM} - \frac{\partial G_j}{\partial t}, \tag{4.5.3}$$

where

$$\mathbf{G} = \varepsilon_0 \mathbf{E} \times \mathbf{B}, \quad G_j = \varepsilon_0 \varepsilon_{jkl} E_k B_l \tag{4.5.4}$$

is the electromagnetic momentum density and

$$T_{ij}^{EM} = P_i E_j' - B_i M_j' + \varepsilon_0 E_i E_j + \frac{1}{\mu_0} B_i B_j$$
$$- \frac{1}{2} \left(\varepsilon_0 E_k E_k + \frac{1}{\mu_0} B_k B_k - 2 M_k' B_k \right) \delta_{ij} \tag{4.5.5}$$

is the electromagnetic stress tensor. In the derivation of Eq. (4.5.3), certain terms on the order or $(v/c)^2$ have been neglected [20]. For later use we introduce the electromagnetic momentum density per unit mass, $\mathbf{g}$, through

$$\mathbf{g} = \mathbf{G}/\rho. \tag{4.5.6}$$

Then, with the use of the conservation of mass, it can be shown from Eq. (4.5.3) that

$$F_j^{EM} = \left(T_{ij}^{EM} + \rho v_i g_j\right)_{,i} - \rho \frac{dg_j}{dt}, \qquad (4.5.7)$$

or

$$\mathbf{F}^{EM} = \nabla \cdot (\mathbf{T}^{EM} + \rho \mathbf{v} \mathbf{g}) - \rho \frac{d\mathbf{g}}{dt}. \qquad (4.5.8)$$

It is also useful to write the electromagnetic body power as the sum of the divergence of a vector field and the time rate of a scalar field, i.e.,

$$W^{EM} = -\nabla \cdot [\mathbf{S} - \mathbf{v}(\mathbf{E} \cdot \mathbf{P})] - \frac{\partial U^F}{\partial t}, \qquad (4.5.9)$$

or

$$W^{EM} + \frac{\partial U^F}{\partial t} = -\nabla \cdot [\mathbf{S} - \mathbf{v}(\mathbf{E} \cdot \mathbf{P})], \qquad (4.5.10)$$

where

$$\mathbf{S} = \mathbf{E} \times \mathbf{H},$$
$$U^F = \frac{\varepsilon_0}{2} \mathbf{E} \cdot \mathbf{E} + \frac{1}{2\mu_0} \mathbf{B} \cdot \mathbf{B}. \qquad (4.5.11)$$

From Eqs. (4.4.4) and (4.5.9), we obtain that

$$\mathbf{E} \cdot \frac{\partial \mathbf{D}}{\partial t} + \mathbf{H} \cdot \frac{\partial \mathbf{B}}{\partial t} = -\nabla \cdot \mathbf{S}, \qquad (4.5.12)$$

which is the same as Eq. (1.10.3) when $\mathbf{J}$ is zero for an insulator. We note that Eq. (1.10.3) is a direct consequence of Maxwell's equations.

We also define the Poynting vector in R_C by

$$\mathbf{S}' = \mathbf{E}' \times \mathbf{H}'. \tag{4.5.13}$$

It can be shown that

$$\mathbf{S}' = \mathbf{S} + \left[\mathbf{T}^{EM} + \mathbf{v}\mathbf{G} \right.$$

$$\left. - \frac{1}{2} \left(\varepsilon_0 \mathbf{E} \cdot \mathbf{E} + \frac{1}{\mu_0} \mathbf{B} \cdot \mathbf{B} + 2\mathbf{E} \cdot \mathbf{P} \right) \mathbf{1} \right] \cdot \mathbf{v}$$

$$= \mathbf{S} + [\mathbf{T}^{EM} + \mathbf{v}\mathbf{G} - (U^F + \mathbf{E} \cdot \mathbf{P})\mathbf{1}] \cdot \mathbf{v}, \tag{4.5.14}$$

where $\mathbf{1}$ is the second-order unit tensor. From Eq. (4.5.9), using Eq. (4.5.14), we obtain

$$W^{EM} = -\rho \frac{d}{dt} \left(\frac{U^F}{\rho} \right) + \nabla \cdot [(\mathbf{T}^{EM} + \mathbf{v}\mathbf{G}) \cdot \mathbf{v} - \mathbf{S}']. \tag{4.5.15}$$

Chapter 5

Electromagnetoelastic Balance Laws

In this chapter, the basic laws of electrodynamics and mechanics in integral forms are applied systematically to the two-continuum model of an electromagnetoelastic solid in the previous chapter. The integral balance laws are reduced to differential forms. They are also used to obtain boundary and discontinuity or jump conditions. The electromagnetic fields in the fixed reference frame (R_G) are $\mathbf{E}$, $\mathbf{B}$, $\mathbf{D}$, $\mathbf{H}$, $\mathbf{M}$ and $\mathbf{P}$. Those in the rest frame (R_C) are $\mathbf{E}'$, $\mathbf{B}'$, $\mathbf{D}'$, $\mathbf{H}'$, $\mathbf{M}'$ and $\mathbf{P}'$. The material forms of the fields are $\mathcal{E}_K$, $\mathcal{B}_K$, $\mathcal{D}_K$, $\mathcal{H}_K$, $\mathcal{M}_K$ and $\mathcal{P}_K$.

5.1 Integral Balance Laws

For electromagnetic fields in insulators, the balance laws in integral forms are

$$\oint_c \mathbf{E} \cdot d\mathbf{y} = -\frac{\partial}{\partial t} \int_s \mathbf{n} \cdot \mathbf{B} ds, \qquad (5.1.1)$$

$$\oint_c \mathbf{H} \cdot d\mathbf{y} = \frac{\partial}{\partial t} \int_s \mathbf{n} \cdot \mathbf{D} ds, \qquad (5.1.2)$$

$$\int_s \mathbf{n} \cdot \mathbf{D} ds = 0, \qquad (5.1.3)$$

$$\int_s \mathbf{n} \cdot \mathbf{B} ds = 0. \qquad (5.1.4)$$

In addition, we have the following relationships:

$$\mathbf{D} = \varepsilon_0 \mathbf{E} + \mathbf{P}, \quad \mathbf{H} = \frac{1}{\mu_0} \mathbf{B} - \mathbf{M}. \tag{5.1.5}$$

Using Eq. (2.5.3), we can write Eqs. (5.1.1) and (5.1.2) as

$$\oint_c \mathbf{E}' \cdot d\mathbf{y} = -\frac{d}{dt} \int_s d\mathbf{s} \cdot \mathbf{B}, \tag{5.1.6}$$

$$\oint_c \mathbf{H}' \cdot d\mathbf{y} = \frac{d}{dt} \int_s d\mathbf{s} \cdot \mathbf{D}, \tag{5.1.7}$$

or

$$\oint_c \mathbf{E}' \cdot d\mathbf{y} = -\int_s d\mathbf{s} \cdot \mathbf{B}^*, \tag{5.1.8}$$

$$\oint_c \mathbf{H}' \cdot d\mathbf{y} = \int_s d\mathbf{s} \cdot \mathbf{D}^*, \tag{5.1.9}$$

where

$$\mathbf{E}' = \mathbf{E} + \mathbf{v} \times \mathbf{B}, \tag{5.1.10}$$
$$\mathbf{H}' = \mathbf{H} - \mathbf{v} \times \mathbf{D}.$$

The conservation of mass of the combined lattice and bound charge continuum is

$$\frac{d}{dt} \int_v \rho \, dv = 0. \tag{5.1.11}$$

The linear momentum equation of the combined continuum can be written as

$$\frac{d}{dt} \int_v \rho \mathbf{v} \, dv = \int_s \mathbf{t} \, ds$$

$$+ \int_v [\rho \mathbf{f} + \mu^l(\mathbf{y}) \mathbf{E}(\mathbf{y}) + \mu^l(\mathbf{y}) \mathbf{v}(\mathbf{y}) \times \mathbf{B}(\mathbf{y})$$

$$+ \mu^b(\mathbf{y} + \boldsymbol{\eta}) \mathbf{E}(\mathbf{y} + \boldsymbol{\eta}) + \mu^b(\mathbf{y} + \boldsymbol{\eta})(\mathbf{v} + \dot{\boldsymbol{\eta}})$$

$$\times \mathbf{B}(\mathbf{y} + \boldsymbol{\eta}) + \mathbf{M}' \cdot (\mathbf{B}\nabla)] \, dv. \tag{5.1.12}$$

With the electromagnetic body force in Eq. (4.2.1), the linear momentum equation can be written as

$$\frac{d}{dt}\int_v \rho\mathbf{v}dv = \int_s \mathbf{t}\,ds + \int_v \left(\rho\mathbf{f} + \mathbf{F}^{EM}\right)dv. \qquad (5.1.13)$$

The linear momentum equation can also be written in another form which will be convenient for boundary conditions. From Eqs. (4.5.3) and (4.5.4), we have

$$F_j^{EM} = T_{ij,i}^{EM} - \frac{\partial G_j}{\partial t}, \qquad (5.1.14)$$

$$\mathbf{G} = \varepsilon_0 \mathbf{E} \times \mathbf{B}, \quad G_j = \varepsilon_0 \varepsilon_{jkl} E_k B_l, \qquad (5.1.15)$$

where $\mathbf{G}$ is the electromagnetic momentum density. Applying the transport theorem in Eq. (2.5.5) to $\mathbf{G}$, we obtain

$$\frac{d}{dt}\int_v \mathbf{G}dv = \int_v \frac{\partial \mathbf{G}}{\partial t}dv + \int_s \mathbf{n}\cdot\mathbf{v}\mathbf{G}ds. \qquad (5.1.16)$$

With Eqs. (5.1.14) and (5.1.16), we can write the linear momentum equation in Eq. (5.1.13) as

$$\frac{d}{dt}\int_v \rho(\mathbf{v} + \mathbf{g})dv$$

$$= \int_v \rho\mathbf{f}dv + \int_s \mathbf{n}\cdot(\boldsymbol{\tau} + \mathbf{T}^{EM} + \mathbf{v}\mathbf{G})ds. \qquad (5.1.17)$$

The angular momentum equation about the origin of $\mathbf{y}$ for the combined continuum takes the following form:

$$\frac{d}{dt}\int_v \mathbf{y} \times \rho\mathbf{v}dv = \int_s \mathbf{y} \times \mathbf{t}\,ds$$

$$+ \int_v \Big\{ \mathbf{y} \times \Big[\rho\mathbf{f} + \mu^l(\mathbf{y})\mathbf{E}(\mathbf{y}) + \mu^l(\mathbf{y})\mathbf{v}(\mathbf{y}) \times \mathbf{B}(\mathbf{y}) + \mathbf{M}'\cdot(\mathbf{B}\nabla)\Big]$$

$$+ (\mathbf{y} + \boldsymbol{\eta})\Big[\mu^b(\mathbf{y} + \boldsymbol{\eta})\mathbf{E}(\mathbf{y} + \boldsymbol{\eta}) + \mu^b(\mathbf{y} + \boldsymbol{\eta})(\mathbf{v} + \dot{\boldsymbol{\eta}}) \times \mathbf{B}(\mathbf{y} + \boldsymbol{\eta})\Big]$$

$$+ \mathbf{M}' \times \mathbf{B}\Big\}dv. \qquad (5.1.18)$$

Using the electromagnetic body couple in Eq. (4.3.1), we can write the angular momentum equation in Eq. (5.1.18) as

$$\frac{d}{dt} \int_v \mathbf{y} \times \rho \mathbf{v} dv = \int_s \mathbf{y} \times \mathbf{t} \, ds$$

$$+ \int_v \left[\mathbf{y} \times \left(\rho \mathbf{f} + \mathbf{F}^{EM} \right) + \mathbf{C}^{EM} \right] dv. \quad (5.1.19)$$

Similar to the derivation of Eq. (5.1.17), using Eq. (4.5.8), we can write the angular momentum equation in Eq. (5.1.19) as

$$\frac{d}{dt} \int_v \rho \mathbf{y} \times (\mathbf{v} + \mathbf{g}) \, dv = \int_v \mathbf{y} \times \rho f dv$$

$$+ \int_s -\mathbf{n} \cdot \left(\boldsymbol{\tau} + \mathbf{T}^{EM} + \mathbf{v} \mathbf{G} \right) \times \mathbf{y} ds. \, (5.1.20)$$

For the conservation of energy of the combined continuum, we have

$$\frac{d}{dt} \int_v \rho \left(\frac{1}{2} \mathbf{v} \cdot \mathbf{v} + \varepsilon \right) dv$$

$$= \int_s \mathbf{t} \cdot \mathbf{v} ds$$

$$+ \int_v \left[\rho \mathbf{f} + \mu^l(\mathbf{y}) \mathbf{E}(\mathbf{y}) + \mu^l(\mathbf{y}) \mathbf{v}(\mathbf{y}) \times \mathbf{B}(\mathbf{y}) \right] \cdot \mathbf{v}$$

$$+ \int_v \left[\mu^b(\mathbf{y} + \boldsymbol{\eta}) \mathbf{E}(\mathbf{y} + \boldsymbol{\eta}) \right.$$

$$\left. + \mu^b(\mathbf{y} + \boldsymbol{\eta})(\mathbf{v} + \dot{\boldsymbol{\eta}}) \times \mathbf{B}(\mathbf{y} + \boldsymbol{\eta}) \right] \cdot (\mathbf{v} + \dot{\boldsymbol{\eta}}) dv$$

$$- \int_v \mathbf{M}' \cdot \frac{\partial \mathbf{B}}{\partial t} dv, \quad (5.1.21)$$

where ε is the internal energy per unit mass. With the use of the electromagnetic body power in Eq. (4.4.1), the energy equation takes the following form:

$$\frac{d}{dt} \int_v \rho \left(\frac{1}{2} \mathbf{v} \cdot \mathbf{v} + \varepsilon \right) dv = \int_s \mathbf{t} \cdot \mathbf{v} ds + \int_v \rho \mathbf{f} \cdot \mathbf{v} dv + \int_v W^{EM} dv.$$

$$(5.1.22)$$

Equation (5.1.22) can also be written as

$$\frac{d}{dt} \int_v \left[\rho \left(\frac{1}{2} \mathbf{v} \cdot \mathbf{v} + \varepsilon \right) + U^F \right] dv = \int_v \rho \mathbf{f} \cdot \mathbf{v} dv$$

$$+ \int_s n_j \left(\tau_{jk} v_k - S_j + v_j P_k E_k + v_j U^F \right) ds, \qquad (5.1.23)$$

where Eq. (4.5.9) has been used, and

$$S_j = \varepsilon_{jkl} E_k H_l,$$

$$U^F = \frac{\varepsilon_0}{2} \mathbf{E} \cdot \mathbf{E} + \frac{1}{2\mu_0} \mathbf{B} \cdot \mathbf{B}. \qquad (5.1.24)$$

From Eq. (5.1.21), using Eq. (4.5.15), we obtain

$$\frac{d}{dt} \int_v \rho \left[\left(\frac{1}{2} \mathbf{v} \cdot \mathbf{v} + \varepsilon \right) + \frac{U^F}{\rho} \right] dv = \int_v \rho \mathbf{f} \cdot \mathbf{v} dv$$

$$+ \int_s \mathbf{n} \cdot \left[(\boldsymbol{\tau} + \mathbf{T}^{EM} + \mathbf{v}\mathbf{G}) \cdot \mathbf{v} - \mathbf{S}' \right] ds. \qquad (5.1.25)$$

5.2 Differential Balance Laws

Equations (5.1.1)–(5.1.4) can be converted to differential forms using Stokes' theorem and the divergence theorem in Appendix C as

$$\nabla \times \mathbf{E} = -\frac{\partial \mathbf{B}}{\partial t}, \qquad (5.2.1)$$

$$\nabla \times \mathbf{H} = \frac{\partial \mathbf{D}}{\partial t}, \qquad (5.2.2)$$

$$\nabla \cdot \mathbf{D} = 0, \qquad (5.2.3)$$

$$\nabla \cdot \mathbf{B} = 0. \qquad (5.2.4)$$

Similarly, from Eqs. (5.1.8) and (5.1.9), we can obtain

$$\nabla \times \mathbf{E}' = -\mathbf{B}^*, \qquad (5.2.5)$$

$$\nabla \times \mathbf{H}' = \mathbf{D}^*. \qquad (5.2.6)$$

The differential form of the conservation of mass is the same as that from Chapter 3:

$$\rho^0 = \rho J \quad \text{or} \quad \dot{\rho} + \rho v_{k,k} = 0. \tag{5.2.7}$$

From the linear momentum equation in Eq. (5.1.13), i.e.,

$$\frac{d}{dt} \int_v \rho \mathbf{v} dv = \int_s \mathbf{t} \, ds + \int_v \left(\rho \mathbf{f} + \mathbf{F}^{EM} \right) = 0, \tag{5.2.8}$$

we obtain

$$\int_v \left(\rho \dot{\mathbf{v}} - \nabla \cdot \boldsymbol{\tau} - \rho \mathbf{f} - \mathbf{F}^{EM} \right) dv = 0, \tag{5.2.9}$$

which yields

$$\rho \frac{d\mathbf{v}}{dt} = \nabla \cdot \boldsymbol{\tau} + \rho \mathbf{f} + \mathbf{F}^{EM}. \tag{5.2.10}$$

Using Eq. (4.5.8), we can write Eq. (5.2.10) as

$$\rho \frac{d}{dt} (\mathbf{v} + \mathbf{g}) = \rho \mathbf{f} + \nabla \cdot (\boldsymbol{\tau} + \mathbf{T}^{EM} + \rho \mathbf{v} \mathbf{g}), \tag{5.2.11}$$

which can also be obtained from Eq. (5.1.17) directly.

From the angular momentum equation in Eq. (5.1.19), i.e.,

$$\frac{d}{dt} \int_v \mathbf{y} \times \rho \mathbf{v} dv = \int_s \mathbf{y} \times \mathbf{t} \, ds$$

$$+ \int_v \left[\mathbf{y} \times (\rho \mathbf{f} + \mathbf{F}^{EM}) + \mathbf{C}^{EM} \right] dv, \tag{5.2.12}$$

we obtain

$$\int_v \mathbf{y} \times \left(\nabla \cdot \boldsymbol{\tau} + \rho \mathbf{f} + \mathbf{F}^{EM} - \rho \dot{\mathbf{v}} \right) dv$$

$$+ \int_v \left(\mathbf{e}_k \varepsilon_{kij} \tau_{ij} + \mathbf{C}^{EM} \right) dv = 0, \tag{5.2.13}$$

which results in

$$\varepsilon_{kij} \tau_{ij} + C_k^{EM} = 0. \tag{5.2.14}$$

Equation (5.2.14) can be written as

$$\varepsilon_{kij}(\tau_{ij} + P_i E_j' + M_i' B_j) = 0, \qquad (5.2.15)$$

which implies that

$$\tau_{ij}^S = \tau_{ij} + P_i E_j' + M_i' B_j \qquad (5.2.16)$$

is symmetric. From the other integral form of the angular momentum equation in Eq. (5.1.20), we can obtain

$$\varepsilon_{kij}\left(\tau_{ij} + T_{ij}^{EM}\right) = 0, \qquad (5.2.17)$$

which is equivalent to Eq. (5.2.14).

From the energy equation in Eq. (5.1.22), i.e.,

$$\frac{d}{dt}\int_v \rho\left(\frac{1}{2}\mathbf{v}\cdot\mathbf{v} + \varepsilon\right) dv = \int_s \mathbf{t}\cdot\mathbf{v}ds + \int_v \rho\mathbf{f}\cdot\mathbf{v}dv$$

$$+ \int_v W^{EM}dv, \qquad (5.2.18)$$

we obtain

$$\int_v (\rho\dot{\mathbf{v}} - \nabla\cdot\boldsymbol{\tau} - \rho\mathbf{f} - \mathbf{F}^{EM})\cdot\mathbf{v}dv$$

$$+ \int_v (\rho\dot{\varepsilon} - \tau_{ij}v_{j,i} - \rho\mathbf{E}'\cdot\dot{\boldsymbol{\pi}} + \mathbf{M}'\cdot\dot{\mathbf{B}})dv = 0, \qquad (5.2.19)$$

where Eq. (4.4.3) has been used. Equation (5.2.19) implies that

$$\rho\frac{d\varepsilon}{dt} = \tau_{ij}v_{j,i} + \rho E_i'\frac{d\pi_i}{dt} - M_i'\frac{dB_i}{dt}. \qquad (5.2.20)$$

5.3   Boundary Conditions

We consider a nonmaterial discontinuity surface [18,20] in general first. A material interface or boundary will be discussed as a special case later. Consider the discontinuity surface in Fig. 5.1. The unit normal **n** points from the "−" side to the "+" side as shown. We use

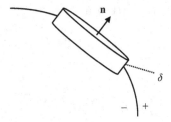

Fig. 5.1.   A discontinuity surface with a pillbox.

square brackets to represent the jump of a physical quantity across
the interface. For example,

$$[\phi] = \phi^+ - \phi^-. \tag{5.3.1}$$

We assume the continuity of **y** and hence the continuity of **v** across
the interface, i.e.,

$$[\mathbf{y}] = 0, \quad [\mathbf{v}] = 0. \tag{5.3.2}$$

Jump conditions across a discontinuity surface can be obtained from
the balance laws in integral form by, e.g., applying them to the pillbox in Fig. 5.1. Since we have treated jump conditions in general
in Section 3.3, we will use the results of Section 3.3 directly. For an
integral balance law in the following general form from Eq. (3.3.1),
i.e.,

$$\frac{d}{dt} \int_v \rho \Psi dv = \int_v \rho \Gamma dv + \int_s \mathbf{n} \cdot \boldsymbol{\Phi} ds, \tag{5.3.3}$$

the corresponding jump conditions across a discontinuity surface is
given by Eq. (3.3.10), i.e.,

$$[\rho \Psi \Theta + \mathbf{n} \cdot \boldsymbol{\Phi}] = 0, \tag{5.3.4}$$

where, from Eqs. (2.6.4) and (2.6.5),

$$\Theta = N - \mathbf{v} \cdot \mathbf{n}, \quad N = \mathbf{v}^s \cdot \mathbf{n}. \tag{5.3.5}$$

$\mathbf{v}^s$ is the velocity of the discontinuity surface. N is the normal velocity of the discontinuity surface. Θ is the relative normal velocity of

the discontinuity surface with respect to the material. For the conservation of mass, we have

$$\frac{d}{dt} \int_v \rho \, dv = 0, \tag{5.3.6}$$

$$\Psi = 1, \quad \Gamma = 0, \quad \Phi = 0, \tag{5.3.7}$$

$$[\rho \Theta] = 0. \tag{5.3.8}$$

For the linear momentum equation, from Eq. (5.1.17), we have

$$\frac{d}{dt} \int_v \rho(\mathbf{v} + \mathbf{g}) dv$$

$$= \int_v \rho \mathbf{f} dv + \int_s \mathbf{n} \cdot (\boldsymbol{\tau} + \mathbf{T}^{EM} + \mathbf{v}\mathbf{G}) ds, \tag{5.3.9}$$

$$\Psi = \mathbf{v} + \mathbf{g}, \quad \Gamma = \mathbf{f},$$

$$\Phi = \boldsymbol{\tau} + \mathbf{T}^{EM} + \mathbf{v}\mathbf{G}, \tag{5.3.10}$$

$$[\rho(\mathbf{v} + \mathbf{g})\Theta + \mathbf{n} \cdot (\boldsymbol{\tau} + \mathbf{T}^{EM} + \mathbf{v}\mathbf{G})] = 0. \tag{5.3.11}$$

For the angular momentum equation, from Eq. (5.1.20), we have

$$\frac{d}{dt} \int_v \rho \mathbf{y} \times (\mathbf{v} + \mathbf{g}) \, dv$$

$$= \int_v \rho \mathbf{y} \times \mathbf{f} dv + \int_s -\mathbf{n} \cdot (\boldsymbol{\tau} + \mathbf{T}^{EM} + \mathbf{v}\mathbf{G}) \times \mathbf{y} ds, \tag{5.3.12}$$

$$\Psi = \mathbf{y} \times (\mathbf{v} + \mathbf{g}), \quad \Gamma = \mathbf{y} \times \mathbf{f},$$

$$\Phi = -(\boldsymbol{\tau} + \mathbf{T}^{EM} + \mathbf{v}\mathbf{G}) \times \mathbf{y}, \tag{5.3.13}$$

$$[\mathbf{y} \times \rho(\mathbf{v} + \mathbf{g})\Theta - \mathbf{n} \cdot (\boldsymbol{\tau} + \mathbf{T}^{EM} + \mathbf{v}\mathbf{G}) \times \mathbf{y}] = 0. \tag{5.3.14}$$

We note that Eq. (5.3.14) is implied by Eq. (5.3.11). For the energy equation, from Eq. (5.1.25), we have

$$\frac{d}{dt} \int_v \rho \left[\left(\frac{1}{2}\mathbf{v} \cdot \mathbf{v} + \varepsilon \right) + \frac{U^F}{\rho} \right] dv = \int_v \rho \mathbf{f} \cdot \mathbf{v} dv$$

$$+ \int_s \mathbf{n} \cdot [(\boldsymbol{\tau} + \mathbf{T}^{EM} + \mathbf{v}\mathbf{G}) \cdot \mathbf{v} - \mathbf{S}'] ds, \tag{5.3.15}$$

$$\Psi = \left(\frac{1}{2}\mathbf{v} \cdot \mathbf{v} + \varepsilon\right) + \frac{U^F}{\rho}, \quad \Gamma = \mathbf{f} \cdot \mathbf{v},$$

$$\Phi = \left(\boldsymbol{\tau} + \mathbf{T}^{EM} + \mathbf{v}\mathbf{G}\right) \cdot \mathbf{v} - \mathbf{S}', \tag{5.3.16}$$

$$[\{\rho\left(\mathbf{v} \cdot \mathbf{v}/2 + \varepsilon\right) + U^F\}\Theta$$
$$+ \mathbf{n} \cdot \{(\boldsymbol{\tau} + \mathbf{T}^{EM} + \mathbf{v}\mathbf{G}) \cdot \mathbf{v} - \mathbf{S}'\}] = 0. \tag{5.3.17}$$

Obviously, for the following two electromagnetic balance laws:

$$\int_s \mathbf{n} \cdot \mathbf{D}ds = 0, \tag{5.3.18}$$

$$\int_s \mathbf{n} \cdot \mathbf{B}ds = 0, \tag{5.3.19}$$

the corresponding jump conditions are

$$\mathbf{n} \cdot [\mathbf{D}] = 0, \tag{5.3.20}$$

$$\mathbf{n} \cdot [\mathbf{B}] = 0. \tag{5.3.21}$$

The two remaining electromagnetic balance laws, i.e.,

$$\oint_c \mathbf{E} \cdot d\mathbf{y} = -\frac{\partial}{\partial t} \int_s d\mathbf{s} \cdot \mathbf{B}, \tag{5.3.22}$$

$$\oint_c \mathbf{H} \cdot d\mathbf{y} = \frac{\partial}{\partial t} \int_s d\mathbf{s} \cdot \mathbf{D}, \tag{5.3.23}$$

are not covered by the general balance law in Eq. (5.3.3). Hence they need a separate treatment. We apply Eq. (5.3.22) to the fixed contour near the discontinuity surface in Fig. 5.2. δ is infinitesimal and the line integral over δ is negligible. The normal velocity of the discontinuity surface is N. During Δt, the discontinuity surface has moved by $N\Delta t < \delta$. The change of the integration over s on the right-hand side of Eq. (5.3.22) has two origins. One is due to the time dependence of the fields in the integrand which is negligible because the relevant fields are assumed to be bounded and s is proportional to δ. The other is due to the changes of the two areas within the contour occupied by $\mathbf{B}^-$ and $\mathbf{B}^+$, respectively, which needs to be

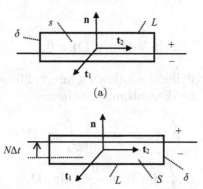

Fig. 5.2. A fixed contour near the discontinuity surface. (a) At time t. (b) At time $t + \Delta t$.

considered. We have [6,9]

$$-E_2^+ L + E_2^- L = -\lim_{\Delta t \to 0} \frac{1}{\Delta t}(B_1^- N \Delta t L - B_1^+ N \Delta t L), \qquad (5.3.24)$$

or

$$-E_2^+ + E_2^- = -(B_1^- N - B_1^+ N), \qquad (5.3.25)$$

$$-[E_2] = N[B_1]. \qquad (5.3.26)$$

If a contour in the plane containing $\mathbf{n}$ and $\mathbf{t}_1$ is used, it leads to

$$[E_1] = N[B_2]. \qquad (5.3.27)$$

Equations (5.3.26) and (5.3.27) can be written together as

$$\varepsilon_{ijk} n_j [E_k] - N[B_i] = 0, \qquad (5.3.28)$$

or

$$[\mathbf{n} \times \mathbf{E} - N\mathbf{B}] = 0. \qquad (5.3.29)$$

Similarly, Eq. (5.3.23) leads to

$$\varepsilon_{ijk} n_j [H_k] + N[D_i] = 0, \qquad (5.3.30)$$

or

$$[\mathbf{n} \times \mathbf{H} + N\mathbf{D}] = 0. \tag{5.3.31}$$

Similarly, by applying the following integral balance laws to material contours near the discontinuity surface:

$$\oint_c \mathbf{E}' \cdot d\mathbf{y} = -\frac{d}{dt} \int_s d\mathbf{s} \cdot \mathbf{B}, \tag{5.3.32}$$

$$\oint_c \mathbf{H}' \cdot d\mathbf{y} = \frac{d}{dt} \int_s d\mathbf{s} \cdot \mathbf{D}, \tag{5.3.33}$$

we obtain

$$\varepsilon_{ijk} n_j [E'_k] - \Theta[B_i] = 0, \tag{5.3.34}$$

$$\varepsilon_{ijk} n_j [H'_k] + \Theta[D_i] = 0, \tag{5.3.35}$$

or

$$[\mathbf{n} \times \mathbf{E}' - \Theta\mathbf{B}] = 0,$$
$$[\mathbf{n} \times \mathbf{H}' + \Theta\mathbf{D}] = 0. \tag{5.3.36}$$

It can be shown that Eqs. (5.3.34) and (5.3.35) are equivalent to Eqs. (5.3.28) and (5.3.30). For example, since

$$\begin{aligned}
\varepsilon_{ijk} n_j E'_k &= \varepsilon_{ijk} n_j (E_k + \varepsilon_{kmn} v_m B_n) \\
&= \varepsilon_{ijk} n_j E_k + n_j \varepsilon_{kij} \varepsilon_{kmn} v_m B_n \\
&= \varepsilon_{ijk} n_j E_k + n_j (\delta_{im} \delta_{jn} - \delta_{in} \delta_{jm}) v_m B_n \\
&= \varepsilon_{ijk} n_j E_k + n_j (v_i B_j - v_j B_i), \tag{5.3.37}
\end{aligned}$$

we have

$$\begin{aligned}
\varepsilon_{ijk} n_j [E'_k] &= \varepsilon_{ijk} n_j [E_k] + [v_i n_j B_j] - [n_j v_j B_i] \\
&= \varepsilon_{ijk} n_j [E_k] - n_j v_j [B_i] = N[B_i] - n_j v_j [B_i] \\
&= (N - n_j v_j)[B_i] = \Theta[B_i]. \tag{5.3.38}
\end{aligned}$$

Hence Eq. (5.3.28) implies Eq. (5.3.34) with the use of Eq. (5.3.21). The relationship between Eqs. (5.3.30) and (5.3.35) is similar.

In the special case of a material discontinuity surface such as a material interface or the boundary surface of a material body, we have $\Theta = 0$. The electromagnetic jump conditions are

$$\mathbf{n} \cdot [\mathbf{D}] = 0, \tag{5.3.39}$$

$$\mathbf{n} \cdot [\mathbf{B}] = 0, \tag{5.3.40}$$

$$\mathbf{n} \times [\mathbf{E}'] = 0, \tag{5.3.41}$$

$$\mathbf{n} \times [\mathbf{H}'] = 0. \tag{5.3.42}$$

For a material discontinuity surface the jump condition from the conservation of mass in Eq. (5.3.8) becomes trivial. The one from the conservation of linear momentum in Eq. (5.3.11) reduces to

$$\mathbf{n} \cdot [\boldsymbol{\tau} + \mathbf{T}^{EM} + \mathbf{vG}] = 0. \tag{5.3.43}$$

The jump condition from the conservation of the angular momentum equation in Eq. (5.3.14) leads to Eq. (5.3.43) too. From Eqs. (5.3.41) and (5.3.42), it can be shown that

$$\mathbf{n} \cdot [\mathbf{S}'] = 0. \tag{5.3.44}$$

Then the jump condition in the energy equation in Eq. (5.3.17) also reduces to Eq. (5.3.43) for a material discontinuity surface.

5.4   Summary

In summary, the differential forms of the balance laws are

$$\nabla \times \mathbf{E} = -\frac{\partial \mathbf{B}}{\partial t}, \tag{5.4.1}$$

$$\nabla \times \mathbf{H} = \frac{\partial \mathbf{D}}{\partial t}, \tag{5.4.2}$$

$$\nabla \cdot \mathbf{D} = 0, \tag{5.4.3}$$

$$\nabla \cdot \mathbf{B} = 0, \tag{5.4.4}$$

$$\rho^0 = \rho J, \qquad (5.4.5)$$

$$\rho \frac{d\mathbf{v}}{dt} = \nabla \cdot \boldsymbol{\tau} + \rho \mathbf{f} + \mathbf{F}^{EM}, \qquad (5.4.6)$$

$$\varepsilon_{kij}\tau_{ij} + C_k^{EM} = 0, \qquad (5.4.7)$$

$$\rho \frac{d\varepsilon}{dt} = \tau_{ij}v_{j,i} + \rho E_i' \frac{d\pi_i}{dt} - M_i' \frac{dB_i}{dt}, \qquad (5.4.8)$$

where

$$\mathbf{F}^{EM} = \mathbf{P} \cdot \nabla \mathbf{E} + \mathbf{M}' \cdot (\mathbf{B}\nabla)$$
$$+ \mathbf{v} \times (\mathbf{P} \cdot \nabla \mathbf{B}) + \rho \dot{\boldsymbol{\pi}} \times \mathbf{B}, \qquad (5.4.9)$$

$$\mathbf{C}^{EM} = \mathbf{P} \times \mathbf{E}' + \mathbf{M}' \times \mathbf{B}, \qquad (5.4.10)$$

$$\mathbf{E}' = \mathbf{E} + \mathbf{v} \times \mathbf{B},$$

$$\mathbf{M}' = \mathbf{M} + \mathbf{v} \times \mathbf{P}. \qquad (5.4.11)$$

In addition,

$$\mathbf{D} = \varepsilon_0 \mathbf{E} + \mathbf{P}, \quad \mathbf{H} = \frac{1}{\mu_0}\mathbf{B} - \mathbf{M}. \qquad (5.4.12)$$

We note that, for insulators, Eqs. (5.4.3) and (5.4.4) are essentially implied by Eqs. (5.4.1) and (5.4.2) except a time derivative. At a discontinuity surface, we have

$$[\mathbf{y}] = 0, \quad [\mathbf{v}] = 0, \qquad (5.4.13)$$

$$[\rho\Theta] = 0, \qquad (5.4.14)$$

$$[\rho(\mathbf{v} + \mathbf{g})\Theta + \mathbf{n} \cdot (\boldsymbol{\tau} + \mathbf{T}^{EM} + \mathbf{v}\mathbf{G})] = 0, \qquad (5.4.15)$$

$$[\{\rho(\mathbf{v} \cdot \mathbf{v}/2 + \varepsilon) + U^F\}\Theta$$
$$+ \mathbf{n} \cdot \{(\boldsymbol{\tau} + \mathbf{T}^{EM} + \mathbf{v}\mathbf{G}) \cdot \mathbf{v} - \mathbf{S}'\}] = 0. \qquad (5.4.16)$$

$$\mathbf{n} \cdot [\mathbf{D}] = 0, \quad \mathbf{n} \cdot [\mathbf{B}] = 0, \qquad (5.4.17)$$

$$[\mathbf{n} \times \mathbf{E}' - \Theta\mathbf{B}] = 0,$$

$$[\mathbf{n} \times \mathbf{H}' + \Theta\mathbf{D}] = 0. \qquad (5.4.18)$$

5.5   Material Form of Balance Laws

Up to this point, most of the differential balance laws in the previous section have been written in terms of the present coordinates y_i in the sense that the spatial derivatives are taken with respect to **y**. Since the reference coordinates of material points are usually known while the present coordinates are not, it is essential to have the equations written in terms of the reference coordinates X_K. For this purpose, for Maxwell's equations, we introduce the following material fields [18,19]:

$$\mathcal{D}_L = JX_{L,i}D_i, \quad \mathcal{B}_L = JX_{L,i}B_i,$$

$$\mathcal{E}_L = E'_k y_{k,L}, \quad \mathcal{H}_L = H'_k y_{k,L}. \tag{5.5.1}$$

Then we can write

$$\int_s D_i ds_i = \int_S D_i JX_{L,i} dS_L$$

$$= \int_S \mathcal{D}_L N_L dS = \int_V \mathcal{D}_{L,L} dV = 0, \tag{5.5.2}$$

which implies that

$$\mathcal{D}_{L,L} = 0. \tag{5.5.3}$$

Similarly, we have

$$\mathcal{B}_{L,L} = 0. \tag{5.5.4}$$

From Eq. (5.1.6),

$$\oint_c \mathbf{E}' \cdot d\mathbf{y} = -\frac{d}{dt}\int_s \mathbf{n} \cdot \mathbf{B} ds \tag{5.5.5}$$

whose component form is

$$\oint_c E'_k dy_k = -\frac{d}{dt}\int_s B_k ds_k. \tag{5.5.6}$$

Then

$$\oint_C \mathcal{E}_L X_{L,k} y_{k,M} dX_M$$

$$= -\frac{d}{dt} \int_S \frac{1}{J} y_{k,L} \mathcal{B}_L J X_{M,k} dS_M, \qquad (5.5.7)$$

$$\oint_C \mathcal{E}_L dX_L = -\frac{d}{dt} \int_S \mathcal{B}_L dS_L, \qquad (5.5.8)$$

$$\oint_C \mathcal{E}_L dX_L = -\int_S \frac{d\mathcal{B}_L}{dt} dS_L. \qquad (5.5.9)$$

With the use of Stokes' theorem,

$$\int_S \varepsilon_{LMN} \mathcal{E}_{N,M} dS_L = -\int_S \frac{d\mathcal{B}_L}{dt} dS_L. \qquad (5.5.10)$$

Hence

$$\varepsilon_{LMN} \mathcal{E}_{N,M} = -\frac{d\mathcal{B}_L}{dt}. \qquad (5.5.11)$$

Similarly, from Eq. (5.1.7),

$$\oint_c \mathbf{H}' \cdot d\mathbf{y} = \frac{d}{dt} \int_s \mathbf{n} \cdot \mathbf{D} ds \qquad (5.5.12)$$

which leads to

$$\varepsilon_{LMN} \mathcal{H}_{N,M} = \frac{d\mathcal{D}_L}{dt}. \qquad (5.5.13)$$

For the linear momentum equation, from Eq. (5.1.17), we have

$$\frac{d}{dt} \int_v \rho(\mathbf{v} + \mathbf{g}) dv = \int_v \rho \mathbf{f} dv$$

$$+ \int_s \mathbf{n} \cdot (\boldsymbol{\tau} + \mathbf{T}^{EM} + \rho \mathbf{vg}) ds. \qquad (5.5.14)$$

The last term on the right-hand side of Eq. (5.3.14) can be written as

$$\int_s n_i(\tau_{ij} + T_{ij}^{EM} + \rho v_i g_j)ds$$

$$= \int_s (\tau_{ij} + T_{ij}^{EM} + \rho v_i g_j)ds_i$$

$$= \int_S (\tau_{ij} + T_{ij}^{EM} + \rho v_i g_j)JX_{L,i}dS_L$$

$$= \int_S (\tau_{ij} + T_{ij}^{EM} + \rho v_i g_j)JX_{L,i}N_L dS$$

$$= \int_V [(\tau_{ij} + T_{ij}^{EM} + \rho v_i g_j)JX_{L,i}]_{,L}dV$$

$$= \int_V K_{Lj,L}dV, \qquad (5.5.15)$$

where we have introduced the first Piola–Kirchhoff stress of

$$K_{Lj} = JX_{L,i}(\tau_{ij} + T_{ij}^{EM} + \rho v_i g_j). \qquad (5.5.16)$$

Substituting Eq. (5.5.15) into Eq. (5.5.14), we obtain

$$\int_V \rho(\dot{v}_j + \dot{g}_j)JdV = \int_V \rho f_j JdV + \int_V K_{Lj,L}dV, \qquad (5.5.17)$$

which implies that

$$K_{Lj,L} + \rho^0 f_j = \rho^0(\dot{v}_j + \dot{g}_j). \qquad (5.5.18)$$

The conservation of angular momentum in Eq. (5.4.7) does not have a spatial derivative. In fact, the angular momentum equation will be satisfied automatically by the constitutive relations in the next chapter. For the energy equation in Eq. (5.4.8), we introduce

$$\chi = \varepsilon - E'_i \pi_i. \tag{5.5.19}$$

Then Eq. (5.4.8) becomes

$$\rho \frac{d\chi}{dt} = \tau_{ij} v_{j,i} - P_j \frac{dE'_j}{dt} - M'_j \frac{dB_j}{dt}. \tag{5.5.20}$$

Let

$$P_{KL} = JX_{K,i}X_{L,j}\tau^S_{ij}$$
$$= JX_{K,i}X_{L,j}(\tau_{ij} + P_i E'_j + M'_i B_j),$$
$$\mathcal{P}_K = JX_{K,i}P_i, \quad \mathcal{M}_K = JX_{K,i}M'_i,$$
$$B_K = B_i y_{i,K}. \tag{5.5.21}$$

Then Eq. (5.5.20) can be written in material form as

$$\rho^0 \frac{d\chi}{dt} = P_{KL}\frac{dE_{KL}}{dt} - \mathcal{P}_K\frac{d\mathcal{E}_K}{dt} - \mathcal{M}_K\frac{dB_K}{dt}. \tag{5.5.22}$$

In summary, corresponding to Eqs. (5.4.1)–(5.4.8), we have

$$\mathcal{D}_{L,L} = 0, \tag{5.5.23}$$

$$\mathcal{B}_{L,L} = 0, \tag{5.5.24}$$

$$\varepsilon_{LMN}\mathcal{E}_{N,M} = -\frac{d\mathcal{B}_L}{dt}, \tag{5.5.25}$$

$$\varepsilon_{LMN}\mathcal{H}_{N,M} = \frac{d\mathcal{D}_L}{dt}, \tag{5.5.26}$$

$$\rho^0 = \rho J, \tag{5.5.27}$$

$$K_{Lj,L} + \rho^0 f_j = \rho^0(\dot{v}_j + \dot{g}_j), \tag{5.5.28}$$

$$\varepsilon_{kij}\tau_{ij} + C^{EM}_k = 0, \tag{5.5.29}$$

$$\rho^0 \frac{d\chi}{dt} = P_{KL}\frac{dE_{KL}}{dt} - \mathcal{P}_K\frac{d\mathcal{E}_K}{dt} - \mathcal{M}_K\frac{dB_K}{dt}, \tag{5.5.30}$$

where all spatial derivatives are taken with respect to $\mathbf{X}$, and

$$\mathcal{D}_L = JX_{L,i}D_i, \quad \mathcal{B}_L = JX_{L,i}B_i,$$

$$\mathcal{E}_L = E'_k y_{k,L}, \quad \mathcal{H}_L = H'_k y_{k,L}, \qquad (5.5.31)$$

$$\mathcal{P}_K = JX_{K,i}P_i, \quad \mathcal{M}_K = JX_{K,i}M'_i,$$

$$B_K = B_i y_{i,K}, \qquad (5.5.32)$$

$$K_{Lj} = JX_{L,i}(\tau_{ij} + T^E_{ij} + \rho v_i g_j),$$

$$P_{KL} = JX_{K,i}X_{L,j}\tau^S_{ij},$$

$$\tau^S_{ij} = \tau_{ij} + P_i E'_j + M'_i B_j. \qquad (5.5.33)$$

For completeness, we derive the material forms of the scalar and vector potentials below [19]. We begin with

$$B_i = \varepsilon_{ijk}A_{k,j}, \qquad (5.5.34)$$

which can be written as

$$J^{-1}y_{i,K}\mathcal{B}_K = \varepsilon_{ijk}A_{k,j}. \qquad (5.5.35)$$

The multiplication of Eq. (5.5.35) by $X_{I,i}$ yields

$$X_{I,i}J^{-1}y_{i,K}\mathcal{B}_K = X_{I,i}\varepsilon_{ijk}A_{k,j}, \qquad (5.5.36)$$

or

$$\begin{aligned}
J^{-1}\mathcal{B}_I &= X_{I,i}\varepsilon_{ijk}A_{k,J}X_{J,j} = X_{I,i}\varepsilon_{ijk}\delta_{lk}A_{l,J}X_{J,j} \\
&= X_{I,i}\varepsilon_{ijk}(y_{l,K}X_{K,k})A_{l,J}X_{J,j} \\
&= (\varepsilon_{ijk}X_{I,i}X_{J,j}X_{K,k})y_{l,K}A_{l,J} = J^{-1}\varepsilon_{IJK}y_{l,K}A_{l,J} \\
&= J^{-1}\varepsilon_{IJK}(y_{l,K}A_l)_{,J}, \qquad (5.5.37)
\end{aligned}$$

where an equation similar to Eq. (2.2.11) has been used. Hence

$$\mathcal{B}_I = \varepsilon_{IJK}(y_{l,K}A_l)_{,J} = \varepsilon_{IJK}\mathcal{A}_{K,J}, \qquad (5.5.38)$$

where the material form of the vector potential $\mathcal{A}_K$ is defined by

$$\mathcal{A}_K = y_{l,K} A_l. \tag{5.5.39}$$

For the scalar potential, since

$$\mathcal{E}_J = E'_j y_{j,J} = (E_j + \varepsilon_{jkl} v_k B_l) y_{j,J}, \tag{5.5.40}$$

we have

$$\mathcal{E}_J X_{J,j} = E_j + \varepsilon_{jkl} v_k B_l. \tag{5.5.41}$$

Hence

$$\begin{aligned} E_j &= \mathcal{E}_J X_{J,j} - \varepsilon_{jkl} v_k B_l \\ &= \mathcal{E}_J X_{J,j} - \varepsilon_{jkl} \left(-y_{k,K} \frac{\partial X_K}{\partial t} \right) B_l, \end{aligned} \tag{5.5.42}$$

where Eq. (2.4.23) has been used. Then

$$\begin{aligned} E_j &= \mathcal{E}_J X_{J,j} - \varepsilon_{jkl} \left(-y_{k,K} \frac{\partial X_K}{\partial t} \right) J^{-1} y_{l,L} \mathcal{B}_L \\ &= \mathcal{E}_J X_{J,j} + J^{-1} \delta_{ij} \varepsilon_{ikl} y_{k,K} y_{l,L} \frac{\partial X_K}{\partial t} \mathcal{B}_L \\ &= \mathcal{E}_J X_{J,j} + J^{-1} (y_{i,J} X_{J,j}) \varepsilon_{ikl} y_{k,K} y_{l,L} \frac{\partial X_K}{\partial t} \mathcal{B}_L \\ &= \mathcal{E}_J X_{J,j} + X_{J,j} J^{-1} \varepsilon_{ikl} y_{i,J} y_{k,K} y_{l,L} \frac{\partial X_K}{\partial t} \mathcal{B}_L \\ &= \mathcal{E}_J X_{J,j} + X_{J,j} \varepsilon_{JKL} \frac{\partial X_K}{\partial t} \mathcal{B}_L, \end{aligned} \tag{5.5.43}$$

where Eq. (2.2.11) has been used. Hence

$$E_j = X_{J,j} \left(\mathcal{E}_J + \varepsilon_{JKL} \frac{\partial X_K}{\partial t} \mathcal{B}_L \right). \tag{5.5.44}$$

Similarly,

$$H_j = X_{J,j}\left(\mathcal{H}_J - \varepsilon_{JKL}\frac{\partial X_K}{\partial t}\mathcal{D}_L\right). \qquad (5.5.45)$$

From Eqs. (5.5.44), (1.9.10)$_2$ and (5.5.39), we have

$$X_{J,j}\left(\mathcal{E}_J + \varepsilon_{JKL}\frac{\partial X_K}{\partial t}\mathcal{B}_L\right)$$

$$= -\varphi_{,j} - \frac{\partial A_j}{\partial t} = -\varphi_{,M}X_{M,j} - \frac{\partial}{\partial t}(X_{K,j}\mathcal{A}_K). \qquad (5.5.46)$$

Multiplying both sides of Eq. (5.4.46) by $y_{j,I}$ and using Eq. (5.5.38), we obtain

$$\mathcal{E}_I + \varepsilon_{IKL}\frac{\partial X_K}{\partial t}\varepsilon_{LMN}\mathcal{A}_{N,M}$$

$$= -\varphi_{,I} - y_{j,I}\left(\frac{\partial^2 X_K}{\partial t\partial y_j}\mathcal{A}_K + X_{K,j}\frac{\partial \mathcal{A}_K}{\partial t}\right). \qquad (5.5.47)$$

With the use of the $\varepsilon - \delta$ identity and Eq. (2.4.23), we can write Eq. (5.5.47) as

$$\mathcal{E}_I = -\varepsilon_{IKL}\frac{\partial X_K}{\partial t}\varepsilon_{LMN}\mathcal{A}_{N,M}$$

$$- \varphi_{,I} - y_{j,I}\left[\mathcal{A}_K\frac{\partial}{\partial y_j}\left(\frac{\partial X_K}{\partial t}\right) + X_{K,j}\frac{\partial \mathcal{A}_K}{\partial t}\right]$$

$$= -(\varphi - v_j A_j)_{,I} - \frac{d\mathcal{A}_I}{dt} = -\mathcal{P}_{,I} - \frac{d\mathcal{A}_I}{dt}, \qquad (5.5.48)$$

where the material scalar potential $\mathcal{P}$ is defined by

$$\mathcal{P} = \varphi - v_j A_j. \qquad (5.5.49)$$

In summary,

$$\mathcal{P} = \varphi - v_j A_j, \quad \mathcal{A}_K = y_{l,K}A_l, \qquad (5.5.50)$$

$$\mathcal{E}_I = -\mathcal{P}_{,I} - \frac{d\mathcal{A}_I}{dt}, \quad \mathcal{B}_I = \varepsilon_{IJK}\mathcal{A}_{K,J}. \qquad (5.5.51)$$

5.6   Quasistatic Balance Laws

The quasistatic approximation of electromagnetic fields (see Section 7.12) is widely used in the theory of electromagnetoelasticity. With this approximation, the terms with time derivatives in Maxwell's equations are dropped. The approximation has a series of consequences which change certain mathematics and physics of the theory fundamentally, but the resulted equations can still describe many acoustic wave dominated phenomena. For quasistatic electromagnetoelasticity, the balance laws in integral forms are

$$\oint_c \mathbf{E} \cdot d\mathbf{y} = 0, \quad \oint_c \mathbf{H} \cdot d\mathbf{y} = 0, \tag{5.6.1}$$

$$\int_s \mathbf{n} \cdot \mathbf{D} ds = 0, \quad \int_s \mathbf{n} \cdot \mathbf{B} ds = 0, \tag{5.6.2}$$

$$\frac{d}{dt} \int_v \rho dv = 0, \tag{5.6.3}$$

$$\frac{d}{dt} \int_v \rho \mathbf{v} dv = \int_s \mathbf{t} ds + \int_v (\rho \mathbf{f} + \mathbf{F}^{EM}) dv, \tag{5.6.4}$$

$$\frac{d}{dt} \int_v \mathbf{y} \times \rho \mathbf{v} dv = \int_s \mathbf{y} \times \mathbf{t} ds$$
$$+ \int_v \left[\mathbf{y} \times (\rho \mathbf{f} + \mathbf{F}^{EM}) + \mathbf{C}^{EM} \right] dv, \tag{5.6.5}$$

$$\frac{d}{dt} \int_v \rho \left(\frac{1}{2} \mathbf{w} \cdot + \varepsilon \right) dv = \int_s \mathbf{t} \cdot \mathbf{v} ds$$
$$+ \int_v \rho \mathbf{f} \cdot \mathbf{v} dv + \int_v W^{EM} dv, \tag{5.6.6}$$

where

$$F_j^{EM} = P_i E_{j,i} + M_i B_{i,j}, \tag{5.6.7}$$

$$C_k^{EM} = \varepsilon_{kij} P_i E_j + \varepsilon_{kij} M_i B_j, \tag{5.6.8}$$

$$W^{EM} = F_j^{EM} v_j + \rho E_j \dot{\pi}_j - M_j \dot{B}_j. \tag{5.6.9}$$

In addition, we have the following relationships:

$$\mathbf{D} = \varepsilon_0 \mathbf{E} + \mathbf{P}, \quad \mathbf{H} = \frac{1}{\mu_0}\mathbf{B} - \mathbf{M}. \tag{5.6.10}$$

Correspondingly, the differential forms of the balance laws are:

$$\nabla \times \mathbf{E} = 0, \quad \nabla \times \mathbf{H} = 0, \tag{5.6.11}$$

$$\nabla \cdot \mathbf{D} = 0, \quad \nabla \cdot \mathbf{B} = 0, \tag{5.6.12}$$

$$\rho^0 = \rho J, \tag{5.6.13}$$

$$\rho \frac{d\mathbf{v}}{dt} = \nabla \cdot \boldsymbol{\tau} + \rho \mathbf{f} + \mathbf{F}^{EM}, \tag{5.6.14}$$

$$\varepsilon_{kij}\tau_{ij} + C_k^{EM} = 0, \tag{5.6.15}$$

$$\rho \frac{d\varepsilon}{dt} = \tau_{ij}v_{j,i} + \rho E_i \frac{d\pi_i}{dt} - M_i \frac{dB_i}{dt}. \tag{5.6.16}$$

Equation (5.6.11) allows the introduction of the following scalar potentials:

$$\mathbf{E} = -\nabla\varphi, \quad \mathbf{H} = -\nabla\psi. \tag{5.6.17}$$

Then constitutive relations with $\mathbf{E}$ and $\mathbf{H}$ as independent variables are desirable.

The electromagnetic body force can be written as

$$F_j^{EM} = T_{ij,i}^{EM}, \tag{5.6.18}$$

where

$$T_{ij}^E = D_i E_j + B_i H_j - \frac{1}{2}\varepsilon_0 E_k E_k \delta_{ij}$$
$$+ \frac{\mu_0}{2}(M_k M_k - H_k H_k)\delta_{ij}. \tag{5.6.19}$$

Then the linear momentum equation in Eq. (5.6.4) can be written as

$$\frac{d}{dt}\int_v \rho\mathbf{v}dv = \int_v \rho\mathbf{f}dv + \int_s \mathbf{n}\cdot(\boldsymbol{\tau} + \mathbf{T}^{EM})ds, \tag{5.6.20}$$

which is convenient for obtaining jump or boundary conditions. From Eq. (4.4.4), it can shown that [17]

$$W^{EM} = -\nabla \cdot [\mathbf{S} - \mathbf{v}(\mathbf{E} \cdot \mathbf{P})] - \frac{\partial U^F}{\partial t}$$

$$= -\nabla \cdot [\mathbf{S} - \mathbf{v}(\mathbf{E} \cdot \mathbf{P}) - \mathbf{v}U^F] - \rho \frac{d}{dt}\left(\frac{U^F}{\rho}\right), \quad (5.6.21)$$

where

$$\mathbf{S} = \varphi \frac{\partial \mathbf{D}}{\partial t} + \psi \frac{\partial \mathbf{B}}{\partial t},$$

$$U^F = \frac{\varepsilon_0}{2}\mathbf{E} \cdot \mathbf{E} + \frac{1}{2\mu_0}\mathbf{B} \cdot \mathbf{B}. \quad (5.6.22)$$

Chapter 6

Electromagnetoelastic Constitutive Relations

The balance laws in the previous chapter are insufficient to form a well-defined initial-boundary-value problem mathematically. Constitutive relations describing the behaviors of the specific material under consideration are necessary. In this chapter, we develop nonlinear electromagnetoelastic constitutive relations [20] which can describe, e.g., piezoelectric, piezomagnetic, magnetoelectric, electrostrictive and magnetostrictive interactions.

6.1 Nonlinear Constitutive Relations

We begin with the energy equation below from Eq. (5.4.8):

$$\rho \frac{d\varepsilon}{dt} = \tau_{ij} v_{j,i} + \rho E'_j \frac{d\pi_j}{dt} - M'_j \frac{dB_j}{dt}. \qquad (6.1.1)$$

Constitutive relations may be written in quite a few different forms with various combinations of independent variables. Some may be relatively more suitable for theoretical study, and others for experimental work. For theoretical analyses, it is often desirable to have $\mathbf{E}'$ and $\mathbf{B}$ as independent constitutive variables because of their direct relations to the scalar and vector potentials. Therefore, we introduce [20]

$$\chi = \varepsilon - E'_i \pi_i. \qquad (6.1.2)$$

Then the energy equation becomes

$$\rho\frac{d\chi}{dt} = \tau_{ij}v_{j,i} - P_j\frac{dE'_j}{dt} - M'_j\frac{dB_j}{dt}, \tag{6.1.3}$$

or

$$\rho\frac{d\chi}{dt} = \tau_{ij}X_{M,i}\frac{d}{dt}(y_{j,M}) - P_j\frac{dE'_j}{dt} - M'_j\frac{dB_j}{dt}, \tag{6.1.4}$$

where we have used

$$v_{j,i} = X_{M,i}\frac{d}{dt}(y_{j,M}). \tag{6.1.5}$$

According to Eq. (6.1.4), we let

$$\chi = \chi(y_{j,M}; E'_i; B_i). \tag{6.1.6}$$

Then

$$\frac{d\chi}{dt} = \frac{\partial\chi}{\partial(y_{j,M})}\frac{d}{dt}(y_{j,M}) + \frac{\partial\chi}{\partial E'_i}\frac{dE'_i}{dt} + \frac{\partial\chi}{\partial B_i}\frac{dB_i}{dt}. \tag{6.1.7}$$

Substituting Eq. (6.1.7) into Eq. (6.1.4), we obtain

$$\rho\frac{\partial\chi}{\partial(y_{j,M})}\frac{d}{dt}(y_{j,M}) + \rho\frac{\partial\chi}{\partial E'_i}\frac{dE'_i}{dt} + \rho\frac{\partial\chi}{\partial B_i}\frac{dB_i}{dt}$$

$$= \tau_{ij}X_{M,i}\frac{d}{dt}(y_{j,M}) - P_j\frac{dE'_j}{dt} - M'_j\frac{dB_j}{dt}, \tag{6.1.8}$$

or

$$\left[X_{M,i}\tau_{ij} - \rho\frac{\partial\chi}{\partial(y_{j,M})}\right]\frac{d}{dt}(y_{j,M})$$

$$- \left(P_j + \rho\frac{\partial\chi}{\partial E'_j}\right)\frac{dE'_j}{dt} - \left(M'_j + \rho\frac{\partial\chi}{\partial B_j}\right)\frac{dB_j}{dt} = 0. \tag{6.1.9}$$

Equation (6.1.9) implies the following constitutive relations:

$$\tau_{ij} = \rho y_{i,M} \frac{\partial \chi}{\partial (y_{j,M})}, \tag{6.1.10}$$

$$P_i = -\rho \frac{\partial \chi}{\partial E_i'}, \tag{6.1.11}$$

$$M_i' = -\rho \frac{\partial \chi}{\partial B_i}. \tag{6.1.12}$$

For rotational invariance (objectivity) [18], χ can be reduced to a function of the following inner products [20]:

$$C_{KL} = y_{i,K} y_{i,L},$$

$$\mathcal{E}_L = y_{i,L} E_i', \quad B_L = y_{i,L} B_i. \tag{6.1.13}$$

Instead of C_{KL}, we will use E_{KL} which is small for infinitesimal deformations:

$$E_{KL} = (C_{KL} - \delta_{KL})/2. \tag{6.1.14}$$

Therefore, we take

$$\chi = \chi(E_{KL}; \mathcal{E}_K; B_K). \tag{6.1.15}$$

Then the constitutive relations in Eqs. (6.1.10)–(6.1.12) become

$$\tau_{ij} = \rho y_{i,M} \frac{\partial \chi}{\partial E_{ML}} y_{j,L}$$

$$+ \rho y_{i,M} \frac{\partial \chi}{\partial \mathcal{E}_M} E_j' + \rho y_{i,M} \frac{\partial \chi}{\partial B_M} B_j, \tag{6.1.16}$$

$$P_i = -\rho y_{i,L} \frac{\partial \chi}{\partial \mathcal{E}_L}, \tag{6.1.17}$$

$$M_i' = -\rho y_{i,L} \frac{\partial \chi}{\partial B_L}. \tag{6.1.18}$$

From Eq. (5.2.16) and the above constitutive relations, we have

$$\tau_{ij}^S = \tau_{ij} + P_i E_j' + M_i' B_j = \rho y_{i,M} \frac{\partial \chi}{\partial E_{ML}} y_{j,L}$$

$$+ \rho y_{i,M} \frac{\partial \chi}{\partial \mathcal{E}_M} E_j' + \rho y_{i,M} \frac{\partial \chi}{\partial B_M} B_j - \rho y_{i,L} \frac{\partial \chi}{\partial \mathcal{E}_L} E_j'$$

$$- \rho y_{i,L} \frac{\partial \chi}{\partial B_L} B_j = \rho y_{i,M} \frac{\partial \chi}{\partial E_{ML}} y_{j,L} = \tau_{ji}^S, \qquad (6.1.19)$$

which is guaranteed to be symmetric. Hence the angular momentum equation in Eq. (5.4.7) is satisfied automatically by the above constitutive relations.

Alternatively, constitutive equations can be obtained from the material form of the energy equation in Eq. (5.5.30), i.e.,

$$\rho^0 \frac{d\chi}{dt} = P_{KL} \frac{dE_{KL}}{dt} - \mathcal{P}_K \frac{d\mathcal{E}_K}{dt} - \mathcal{M}_K \frac{dB_K}{dt}, \qquad (6.1.20)$$

where

$$P_{KL} = JX_{K,i}X_{L,j}\tau_{ij}^S,$$

$$\mathcal{P}_K = JX_{K,i}P_i, \quad \mathcal{M}_K = JX_{K,i}M_i', \qquad (6.1.21)$$

$$\mathcal{E}_K = E_i' y_{i,K}, \quad B_K = B_i y_{i,K}.$$

Equation (6.1.20) suggests the following expression of χ directly:

$$\chi = \chi(E_{KL}; \mathcal{E}_K; B_K). \qquad (6.1.22)$$

From Eqs. (6.1.20) and (6.1.22) we obtain the constitutive relations as

$$P_{KL} = \rho^0 \frac{\partial \chi}{\partial E_{KL}} = P_{LK},$$

$$\qquad (6.1.23)$$

$$\mathcal{P}_K = -\rho^0 \frac{\partial \chi}{\partial \mathcal{E}_K}, \quad \mathcal{M}_K = -\rho^0 \frac{\partial \chi}{\partial B_K},$$

which are equivalent to Eqs. (6.1.16)–(6.1.18). Equation (6.1.23) ensures the symmetry of τ_{kl}^S rather directly.

We note that with the above constitutive relations, the energy equation in Eq. (5.4.8) is already satisfied, in addition to that the angular momentum equation in Eq. (5.4.7) is no longer needed. Using these constitutive relations, we can write the balance laws in Eqs. (5.4.1)–(5.4.6) as equations for $\mathbf{y}$, $\mathbf{E}$, $\mathbf{B}$ and ρ. If $\mathbf{E}$ and $\mathbf{B}$ are further expressed by the scalar and vector potentials through

$$\mathbf{B} = \nabla \times \mathbf{A},$$
$$\mathbf{E} = -\nabla \varphi - \frac{\partial \mathbf{A}}{\partial t}, \tag{6.1.24}$$

then the relevant balance laws can be written as equations for $\mathbf{y}$, φ, $\mathbf{A}$ and ρ.

As an example, χ may be chosen as

$$\chi = \frac{1}{2}c_{KLMN}E_{KL}E_{MN} - \frac{1}{2}\chi^E_{KL}\mathcal{E}_K\mathcal{E}_L - \frac{1}{2}\chi^B_{KL}B_K B_L$$
$$- e_{KLM}\mathcal{E}_K E_{LM} - h_{KLM}B_K E_{LM} - m_{KL}\mathcal{E}_K B_L$$
$$- b^E_{KLMN}\mathcal{E}_K\mathcal{E}_L E_{MN} - b^B_{KLMN}B_K B_L E_{MN} \cdots, \tag{6.1.25}$$

where

$$c_{KLMN} - \text{elastic constants,}$$
$$\chi^E_{KL} - \text{electric susceptibility,}$$
$$\chi^B_{KL} - \text{magnetic susceptibility,}$$
$$e_{KLM} - \text{piezoelectric constants,}$$
$$h_{KLM} - \text{piezomagnetic constants,}$$
$$m_{KL} - \text{magnetoelectric constants,}$$
$$b^E_{KLMN} - \text{electrostrictive constants,}$$
$$b^B_{KLMN} - \text{magnetostrictive constants.}$$

6.2 Rigid Materials

For rigid materials, the strain $E_{KL} = 0$ so that

$$\chi = \chi(\mathcal{E}_K; B_K), \tag{6.2.1}$$

$$\mathcal{P}_K = -\rho^0 \frac{\partial \chi}{\partial \mathcal{E}_K}, \quad \mathcal{M}_K = -\rho^0 \frac{\partial \chi}{\partial B_K}. \tag{6.2.2}$$

When the material is isotropic, it is convenient to use the representation based on the invariants of $\mathcal{E}_K$ and B_K [18], i.e.,

$$\chi = \chi(I_4, I_5, I_{10}), \tag{6.2.3}$$

where

$$I_4 = \mathcal{E}_K \mathcal{E}_K, \quad I_5 = B_K B_K, \quad I_{10} = (\mathcal{E}_K B_K)^2. \tag{6.2.4}$$

Then

$$\begin{aligned}
\mathcal{P}_K &= -\rho^0 \frac{\partial \chi}{\partial \mathcal{E}_K} = -\rho^0 \frac{\partial \chi}{\partial I_4} \frac{\partial I_4}{\partial \mathcal{E}_K} - \rho^0 \frac{\partial \chi}{\partial I_{10}} \frac{\partial I_{10}}{\partial \mathcal{E}_K} \\
&= -\rho^0 \frac{\partial \chi}{\partial I_4} 2\mathcal{E}_K - \rho^0 \frac{\partial \chi}{\partial I_{10}} (2\mathcal{E}_L B_L) B_K \\
&= \chi^E \mathcal{E}_K + m B_K,
\end{aligned} \tag{6.2.5}$$

$$\begin{aligned}
\mathcal{M}_K &= -\rho^0 \frac{\partial \chi}{\partial B_K} = -\rho^0 \frac{\partial \chi}{\partial I_5} \frac{\partial I_5}{\partial B_K} - \rho^0 \frac{\partial \chi}{\partial I_{10}} \frac{\partial I_{10}}{\partial B_K} \\
&= -\rho^0 \frac{\partial \chi}{\partial I_5} 2B_K - \rho^0 \frac{\partial \chi}{\partial I_{10}} (2\mathcal{E}_L B_L) \mathcal{E}_K \\
&= \chi^B B_K + m \mathcal{E}_K,
\end{aligned} \tag{6.2.6}$$

where

$$\chi^E = -2\rho^0 \frac{\partial \chi}{\partial I_4}, \quad \chi^B = -2\rho^0 \frac{\partial \chi}{\partial I_5},$$

$$m = -2\rho^0 \frac{\partial \chi}{\partial I_{10}} \mathcal{E}_L B_L. \tag{6.2.7}$$

For linear and anisotropic materials,

$$\rho^0 \chi = -\frac{1}{2}\chi_{KL}^E \mathcal{E}_K \mathcal{E}_L - \frac{1}{2}\chi_{KL}^B B_K B_L - m_{KL}\mathcal{E}_K B_L. \tag{6.2.8}$$

Then

$$\mathcal{P}_K = -\rho^0 \frac{\partial \chi}{\partial \mathcal{E}_K} = \chi_{KL}^E \mathcal{E}_L + m_{KL} B_L, \tag{6.2.9}$$

$$\mathcal{M}_K = -\rho^0 \frac{\partial \chi}{\partial B_K} = \chi_{KL}^B B_L + m_{LK}\mathcal{E}_L. \tag{6.2.10}$$

If the material is linear and isotropic,

$$\rho^0 \chi = -\frac{1}{2}\chi^E \mathcal{E}_K \mathcal{E}_K - \frac{1}{2}\chi^B B_K B_K - m\mathcal{E}_K B_K, \tag{6.2.11}$$

$$\mathcal{P}_K = -\rho^0 \frac{\partial \chi}{\partial \mathcal{E}_K} = \chi^E \mathcal{E}_K + m B_K, \tag{6.2.12}$$

$$\mathcal{M}_K = -\rho^0 \frac{\partial \chi}{\partial B_K} = \chi^B B_K + m\mathcal{E}_K. \tag{6.2.13}$$

6.3 Elastic Dielectrics

For non-magnetizable electromagnetic dielectrics, χ is independent of **B**. Hence we have

$$\chi = \chi(E_{KL}; \mathcal{E}_K), \tag{6.3.1}$$

and

$$P_{KL} = \rho^0 \frac{\partial \chi}{\partial E_{KL}},$$

$$\mathcal{P}_K = -\rho^0 \frac{\partial \chi}{\partial \mathcal{E}_K}, \quad \mathcal{M}_K = 0. \tag{6.3.2}$$

When the material is isotropic, χ can be written as [18]

$$\chi = \chi(I_1, I_2, I_3, I_4, I_6, I_8), \tag{6.3.3}$$

where, with $\mathbf{E}$ for the strain tensor,

$$I_1 = \text{tr}\mathbf{E} = E_{KK},$$
$$I_2 = \text{tr}\mathbf{E}^2 = E_{KL}E_{LK}, \tag{6.3.4}$$
$$I_3 = \text{tr}\mathbf{E}^3 = E_{KL}E_{LM}E_{MK},$$
$$I_4 = \mathcal{E} \cdot \mathcal{E} = \mathcal{E}_K \mathcal{E}_K,$$
$$I_6 = \mathcal{E} \cdot \mathbf{E} \cdot \mathcal{E} = \mathcal{E}_K E_{KL}\mathcal{E}_L, \tag{6.3.5}$$
$$I_8 = \mathcal{E} \cdot \mathbf{E}^2 \cdot \mathcal{E} = \mathcal{E}_K E_{KL}E_{LM}\mathcal{E}_M,$$
$$\mathbf{E} = E_{KL}\mathbf{I}_K\mathbf{I}_L, \quad \mathcal{E} = \mathcal{E}_K \mathbf{I}_K. \tag{6.3.6}$$

Hence the constitutive relations are

$$P_{KL}\mathbf{I}_K\mathbf{I}_L = \rho^0 \frac{\partial \chi}{\partial I_1}\mathbf{1} + 2\rho^0 \frac{\partial \chi}{\partial I_2}\mathbf{E} + 3\rho^0 \frac{\partial \chi}{\partial I_3}\mathbf{E}^2$$
$$+ \rho^0 \frac{\partial \chi}{\partial I_6}\mathcal{E} \otimes \mathcal{E} + \rho^0 \frac{\partial \chi}{\partial I_8}[\mathcal{E} \otimes (\mathbf{E} \cdot \mathcal{E}) + (\mathbf{E} \cdot \mathcal{E}) \otimes \mathcal{E}],$$
$$\tag{6.3.7}$$

$$\mathcal{P}_L\mathbf{I}_L = -2\rho^0 \frac{\partial \chi}{\partial I_4}\mathcal{E} - 2\rho^0 \frac{\partial \chi}{\partial I_6}\mathbf{E} \cdot \mathcal{E} - 2\rho^0 \frac{\partial \chi}{\partial I_8}\mathbf{E}^2 \cdot \mathcal{E}, \tag{6.3.8}$$

where

$$\mathbf{A} \otimes \mathbf{B} = A_K\mathbf{I}_K B_L\mathbf{I}_L = A_K B_L\mathbf{I}_K\mathbf{I}_L,$$
$$(\mathbf{A} \otimes \mathbf{B})_{KL} = A_K B_L. \tag{6.3.9}$$

For linear materials,

$$\rho^0 \chi = \frac{1}{2}c_{IJKL}E_{IJ}E_{KL} - \frac{1}{2}\chi_{KL}^E \mathcal{E}_K \mathcal{E}_L - e_{KLM}\mathcal{E}_K E_{LM} \tag{6.3.10}$$

which produces

$$P_{KL} = \rho^0 \frac{\partial \chi}{\partial E_{KL}} = c_{KLMN}E_{MN} - e_{MKL}\mathcal{E}_M, \tag{6.3.11}$$

$$\mathcal{P}_K = \chi_{KL}^E \mathcal{E}_L + e_{KLM}E_{LM}. \tag{6.3.12}$$

6.4 Magnetoelasticity

For materials that are not polarizable, χ does not depend on the electric field. We have [18]

$$\chi = \chi(I_1, I_2, I_3, I_5, I_7, I_9), \tag{6.4.1}$$

where

$$
\begin{aligned}
I_5 &= \mathbf{B} \cdot \mathbf{B} = B_K B_K, \\
I_7 &= \mathbf{B} \cdot \mathbf{E} \cdot \mathbf{B} = B_K E_{KL} B_L, \\
I_9 &= \mathbf{B} \cdot \mathbf{E}^2 \cdot \mathbf{B} = B_K E_{KL} E_{LM} B_M.
\end{aligned}
\tag{6.4.2}
$$

6.5 Linear Materials

For linear electromagnetoelastic materials, from Eq. (6.1.25), keeping quadratic terms only, we have [18]

$$
\begin{aligned}
\rho^0 \chi = {} & \frac{1}{2} c_{IJKL} E_{IJ} E_{KL} - \frac{1}{2} \chi_{KL}^E \mathcal{E}_K \mathcal{E}_L - \frac{1}{2} \chi_{KL}^B B_K B_L \\
& - e_{KLM} \mathcal{E}_K E_{LM} - h_{KLM} B_K E_{LM} - m_{KL} \mathcal{E}_K B_L.
\end{aligned}
\tag{6.5.1}
$$

Then

$$P_{KL} = c_{KLMN} E_{MN} - e_{MKL} \mathcal{E}_M - h_{MKL} B_M, \tag{6.5.2}$$

$$\mathcal{P}_K = \chi_{KL}^E \mathcal{E}_L + e_{KLM} E_{LM} + m_{KL} B_L, \tag{6.5.3}$$

$$\mathcal{M}_K = \chi_{KL}^B B_L + h_{KLM} E_{LM} + m_{LK} \mathcal{E}_L. \tag{6.5.4}$$

We note that although Eqs. (6.5.2)–(6.5.4) are linear formally, non-linearities still exist in

$$
\begin{aligned}
E_{KL} &= (y_{i,K} y_{i,L} - \delta_{KL})/2 \\
&= (u_{K,L} + u_{L,K} + u_{M,K} u_{M,L})/2 = E_{LK},
\end{aligned}
\tag{6.5.5}
$$

$$
\begin{aligned}
\mathbf{E}' &\cong \mathbf{E} + \mathbf{v} \times \mathbf{B}, \\
\mathbf{M}' &\cong \mathbf{M} + \mathbf{v} \times \mathbf{P},
\end{aligned}
\tag{6.5.6}
$$

which are implicit in Eqs. (6.5.2)–(6.5.4). These nonlinearities may be neglected in certain applications for a completely linear theory.

6.6   Small Fields on a Finite Bias

Similar to Section 3.11, we consider the following three states of an electromagnetoelastic body:

(i) In the reference state, the body is undeformed and is free of any loads and fields. The mass density is ρ^0.

(ii) Under some finite and static loads, the body assumes its initial or biasing state with finite and static deformations and fields called biasing fields. The fields of the initial state are indicated by a superscript "1". Those that are relevant to the constitutive relations are

$$E^1_{KL}, \quad \mathcal{E}^1_K, \quad B^1_K, \tag{6.6.1}$$

$$P^1_{KL} = \rho^0 \frac{\partial \chi}{\partial E_{KL}}\bigg|_{\mathbf{E}^1, \mathcal{E}^1, \mathbf{B}^1}, \tag{6.6.2}$$

$$\mathcal{P}^1_K = -\rho^0 \frac{\partial \chi}{\partial \mathcal{E}_K}\bigg|_{\mathbf{E}^1, \mathcal{E}^1, \mathbf{B}^1}, \tag{6.6.3}$$

$$\mathcal{M}^1_K = -\rho^0 \frac{\partial \chi}{\partial B_K}\bigg|_{\mathbf{E}^1, \mathcal{E}^1, \mathbf{B}^1}. \tag{6.6.4}$$

(iii) In the present state, the body is under the action of time-dependent loads. The present or final fields are

$$E_{KL} \cong E^1_{KL} + \tilde{E}_{KL}, \quad P_{KL} \cong P^1_{KL} + \tilde{P}_{KL},$$
$$\mathcal{E}_K = \mathcal{E}^1_K + \tilde{\mathcal{E}}_K, \quad \mathcal{P}_K = \mathcal{P}^1_K + \tilde{\mathcal{P}}_K, \tag{6.6.5}$$
$$B_K = B^1_K + \tilde{B}_K, \quad \mathcal{M}_K = \mathcal{M}^1_K + \tilde{\mathcal{M}}_K,$$

where the incremental fields are indicated by a superimposed tilde "~" and are assumed to be infinitesimal. Then, from

$$P_{KL} = \rho^0 \left.\frac{\partial \chi}{\partial E_{KL}}\right|_{\mathbf{E},\mathcal{E},\mathbf{B}},$$

$$\mathcal{P}_K = -\rho^0 \left.\frac{\partial \chi}{\partial \mathcal{E}_K}\right|_{\mathbf{E},\mathcal{E},\mathbf{B}}, \quad \mathcal{M}_K = -\rho^0 \left.\frac{\partial \chi}{\partial B_K}\right|_{\mathbf{E},\mathcal{E},\mathbf{B}}, \tag{6.6.6}$$

it can be shown by Taylor's expansions that the effective constitutive relations for the incremental fields are

$$\tilde{P}_{KL} = \rho^0 \left.\frac{\partial^2 \chi}{\partial E_{KL}\partial E_{MN}}\right|_{\mathbf{E}^1,\mathcal{E}^1,\mathbf{B}^1} \tilde{E}_{MN}$$

$$+ \rho^0 \left.\frac{\partial^2 \chi}{\partial E_{KL}\partial \mathcal{E}_M}\right|_{\mathbf{E}^1,\mathcal{E}^1,\mathbf{B}^1} \tilde{\mathcal{E}}_M + \rho^0 \left.\frac{\partial^2 \chi}{\partial E_{KL}\partial B_M}\right|_{\mathbf{E}^1,\mathcal{E}^1,\mathbf{B}^1} \tilde{B}_M, \tag{6.6.7}$$

$$\tilde{\mathcal{P}}_K = -\rho^0 \left.\frac{\partial^2 \chi}{\partial \mathcal{E}_K\partial E_{MN}}\right|_{\mathbf{E}^1,\mathcal{E}^1,\mathbf{B}^1} \tilde{E}_{MN}$$

$$- \rho^0 \left.\frac{\partial^2 \chi}{\partial \mathcal{E}_K\partial \mathcal{E}_L}\right|_{\mathbf{E}^1,\mathcal{E}^1,\mathbf{B}^1} \tilde{\mathcal{E}}_L - \rho^0 \left.\frac{\partial^2 \chi}{\partial \mathcal{E}_K\partial B_L}\right|_{\mathbf{E}^1,\mathcal{E}^1,\mathbf{B}^1} \tilde{B}_L, \tag{6.6.8}$$

$$\tilde{\mathcal{M}}_K = -\rho^0 \left.\frac{\partial^2 \chi}{\partial B_K\partial E_{MN}}\right|_{\mathbf{E}^1,\mathcal{E}^1,\mathbf{B}^1} \tilde{E}_{MN}$$

$$- \rho^0 \left.\frac{\partial^2 \chi}{\partial B_K\partial \mathcal{E}_L}\right|_{\mathbf{E}^1,\mathcal{E}^1,\mathbf{B}^1} \tilde{\mathcal{E}}_L - \rho^0 \left.\frac{\partial^2 \chi}{\partial B_K\partial B_L}\right|_{\mathbf{E}^1,\mathcal{E}^1,\mathbf{B}^1} \tilde{B}_L. \tag{6.6.9}$$

Equations (6.6.7)–(6.6.9) may be written as

$$\tilde{P}_{KL} = c^1_{KLMN}\tilde{E}_{MN} - e^1_{MKL}\tilde{\mathcal{E}}_M - h^1_{MKL}\tilde{B}_M, \tag{6.6.10}$$

$$\tilde{\mathcal{P}}_K = e^1_{KMN}\tilde{E}_{MN} + \chi^{1E}_{KL}\tilde{\mathcal{E}}_L + m^1_{KL}\tilde{B}_L, \tag{6.6.11}$$

$$\tilde{\mathcal{M}}_K = h^1_{KLM}\tilde{E}_{MN} + m^1_{LK}\tilde{\mathcal{E}}_L + \chi^{1B}_{KL}\tilde{B}_L, \tag{6.6.12}$$

where the effective material constants are defined by

$$c^1_{KLMN} = \rho^0 \frac{\partial^2 \chi}{\partial E_{KL} \partial E_{MN}}\bigg|_{\mathbf{E}^1, \mathcal{E}^1, \mathbf{B}^1}, \tag{6.6.13}$$

$$e^1_{MKL} = -\rho^0 \frac{\partial^2 \chi}{\partial E_{KL} \partial \mathcal{E}_M}\bigg|_{\mathbf{E}^1, \mathcal{E}^1, \mathbf{B}^1},$$

$$h^1_{MKL} = -\rho^0 \frac{\partial^2 \chi}{\partial E_{KL} \partial B_M}\bigg|_{\mathbf{E}^1, \mathcal{E}^1, \mathbf{B}^1}, \tag{6.6.14}$$

$$\chi^{1E}_{KL} = -\rho^0 \frac{\partial^2 \chi}{\partial \mathcal{E}_K \partial \mathcal{E}_L}\bigg|_{\mathbf{E}^1, \mathcal{E}^1, \mathbf{B}^1},$$

$$\chi^{1B}_{KL} = -\rho^0 \frac{\partial^2 \chi}{\partial B_K \partial B_L}\bigg|_{\mathbf{E}^1, \mathcal{E}^1, \mathbf{B}^1}, \tag{6.6.15}$$

$$m^1_{KL} = -\rho^0 \frac{\partial^2 \chi}{\partial \mathcal{E}_K \partial B_L}\bigg|_{\mathbf{E}^1, \mathcal{E}^1, \mathbf{B}^1}. \tag{6.6.16}$$

These effective material constants show plenty of nonlinear coupling effects. For example, the dependence of c^1_{KLMN} on the initial deformation, initial electric field and initial magnetic field are referred to as acoustoelastic, electroacoustic and magnetoacoustic effects [18,19]. Similarly, the dependence of χ^{1E}_{KL} on the initial deformation, electric field and magnetic field describe photoelastic, electro-optical and magneto-optical effects [18,19].

Chapter 7

Linear Problems

This chapter is on linear fields in piezoelectric and nonmagnetizable materials, e.g., [23–25]. Fully electrodynamic couplings are considered. Such a theory has been called piezoelectromagnetism by some researchers. It shows the basic effects of elastic-electrodynamic couplings in a relatively simple manner. The results are useful in acoustic and optic wave devices and the interactions between the two types of waves. At the end of the chapter, the theory of piezoelectromagnetism is reduced to the widely used quasistatic theory of piezoelectricity by a perturbation procedure [19].

7.1 Governing Equations

The three-dimensional equations of linear piezoelectromagnetism consist of Newton's law and Maxwell's equations

$$T_{ji,j} + f_i = \rho \ddot{u}_i \qquad (7.1.1)$$

$$\varepsilon_{ijk} H_{k,j} = \dot{D}_i,$$
$$D_{i,i} = 0, \qquad (7.1.2)$$

$$\varepsilon_{ijk} E_{k,j} = -\dot{B}_i,$$
$$B_{i,i} = 0, \qquad (7.1.3)$$

as well as the following constitutive relations:

$$T_{ij} = c_{ijkl}S_{kl} - e_{kij}E_k,$$

$$D_i = e_{ijk}S_{jk} + \varepsilon_{ij}E_j, \qquad (7.1.4)$$

$$B_i = \mu_0 H_j.$$

In Eq. (7.1.4), the difference between $\mathbf{E}'$ and $\mathbf{E}$ has been neglected and the equation is completely linear. With the introduction of the vector potential, $\mathbf{A}$, and scalar potential, φ, for the electromagnetic fields by

$$E_k = -\varphi_{,k} - \dot{A}_k, \quad B_k = \varepsilon_{kij}A_{j,i}, \qquad (7.1.5)$$

Eq. (7.1.3) is identically satisfied. Then Eqs. (7.1.1), (7.1.2) and (7.1.4) can be written as equations in terms of $\mathbf{u}$, φ and $\mathbf{A}$. Equations (7.1.1)–(7.1.4) can also be reduced to the following two equations for $\mathbf{u}$ and $\mathbf{E}$:

$$c_{ijkl}u_{k,li} = \rho\ddot{u}_j + e_{kij}E_{k,i},$$

$$E_{i,kk} - E_{k,ki} = \mu_0\varepsilon_{ik}\ddot{E}_k + \mu_0 e_{ikl}\ddot{u}_{k,l}, \qquad (7.1.6)$$

which shows the piezoelectric coupling between acoustic and electromagnetic waves.

On the boundary surface of a finite body, the displacement $\mathbf{u}$ or traction $\mathbf{t}$ may be prescribed as mechanical boundary conditions. The boundary conditions for the electromagnetic fields are usually in terms of the tangential components of $\mathbf{E}$ and $\mathbf{H}$ as well as the normal components of $\mathbf{B}$ and $\mathbf{D}$. Consider a finite body occupying a region V. The boundary surface of V is denoted by S with an outward unit normal $\mathbf{n}$. For mechanical boundary conditions, S is partitioned into S_u and S_T. For electromagnetic boundary conditions, we consider the following partitions of S:

$$S_\varphi \cup S_D = S_A \cup S_H = S,$$

$$S_\varphi \cap S_D = S_A \cap S_H = \emptyset. \qquad (7.1.7)$$

On S, we may prescribe

$$u_i = \overline{u}_i \quad \text{on} \quad S_u,$$

$$n_i T_{ij} = \overline{t}_j \quad \text{on} \quad S_T, \tag{7.1.8}$$

$$\varphi = \overline{\varphi} \quad \text{on} \quad S_\varphi,$$

$$D_i n_i + \overline{d} = 0 \quad \text{on} \quad S_D,$$

$$\varepsilon_{ijk} n_j A_k = \overline{a}_i \quad \text{on} \quad S_A, \tag{7.1.9}$$

$$\varepsilon_{ijk} n_j H_k = \overline{h}_i \quad \text{on} \quad S_H, \tag{7.1.10}$$

where $\overline{u}_i$, $\overline{t}_i$, $\overline{\varphi}$, $\overline{d}$, $\overline{a}_i$ and $\overline{h}_i$ are known boundary data.

7.2 Variational Formulation

Consider the following Lagrangian density L and functional Π [26]:

$$L = \frac{1}{2}\dot{u}_k \dot{u}_k - \frac{1}{2}c_{ijkl}S_{ij}S_{kl} + e_{ijk}E_i S_{jk}$$

$$+ \frac{1}{2}\varepsilon_{ij}E_i E_j - \frac{1}{2\mu_0}B_j B_j, \tag{7.2.1}$$

$$\Pi(\mathbf{u}, \mathbf{A}, \varphi) = \int_{t_0}^{t_1} dt \int_V (L + f_k u_k)dV$$

$$+ \int_{t_0}^{t_1} dt \int_{S_T} \overline{t}_k u_k dS - \int_{t_0}^{t_1} dt \int_{S_D} \overline{d}\varphi dS$$

$$- \int_{t_0}^{t_1} dt \int_{S_H} \overline{h}_i A_i dS. \tag{7.2.2}$$

The admissible functions of Π satisfy

$$u_i = \overline{u}_i \quad \text{on} \quad S_u,$$

$$\varphi = \overline{\varphi} \quad \text{on} \quad S_\varphi, \tag{7.2.3}$$

$$\varepsilon_{ijk} n_j A_k = \overline{a}_i \quad \text{on} \quad S_A,$$

$$\delta u_i(\mathbf{x}, t_0) = 0, \quad \delta u_i(\mathbf{x}, t_1) = 0 \quad \text{in} \quad V,$$

$$\delta A_i(\mathbf{x}, t_0) = 0, \quad \delta A_i(\mathbf{x}, t_1) = 0 \quad \text{in} \quad V. \tag{7.2.4}$$

Then,

$$\begin{aligned}
\delta\Pi = &\int_{t_0}^{t_1} dt \int_V [(T_{lk,l} + f_k - \rho\ddot{u}_k)\delta u_k \\
&+ D_{i,i}\delta\varphi + (\dot{D}_i - \varepsilon_{ijk}H_{k,j})\delta A_i]dV \\
&- \int_{t_0}^{t_1} dt \int_{S_T} (n_i T_{ij} - \bar{t}_j)\delta u_j dS \\
&- \int_{t_0}^{t_1} dt \int_{S_D} (D_i n_i + \bar{d})\delta\varphi dS \\
&+ \int_{t_0}^{t_1} dt \int_{S_H} (\varepsilon_{ijk}n_j H_k - \bar{h}_i)\delta A_i dS. \tag{7.2.5}
\end{aligned}$$

Therefore, the stationary condition of Π yields

$$T_{ji,j} + f_i = \rho\ddot{u}_i, \tag{7.2.6}$$

$$\varepsilon_{ijk}H_{k,j} = \dot{D}_i,$$

$$D_{i,i} = 0, \tag{7.2.7}$$

and

$$n_i T_{ij} = \bar{t}_j \quad \text{on} \quad S_T,$$

$$D_i n_i + \bar{d} = 0 \quad \text{on} \quad S_D,$$

$$\varepsilon_{ijk}n_j H_k = \bar{h}_i \quad \text{on} \quad S_H. \tag{7.2.8}$$

7.3 Antiplane Motions of Hexagonal Crystals

The equations of piezoelectromagnetism may be simple or rather complicated depending on material anisotropy. In the rest of this chapter except the last two sections, we study antiplane or shear-horizontal (SH) motions of hexagonal crystals of class (6mm) which are relatively simple mathematically and can show some of the basic physics. The material matrices of polarized ceramics are the same as those of hexagonal crystals. Therefore, the equations for hexagonal crystals are also applicable to polarized ceramics. For antiplane problems of these materials, the equations of piezoelectromagnetism allow the following fields:

$$u_1 = u_2 = 0, \quad u_3 = u_3(x_1, x_2, t),$$
$$E_1 = E_1(x_1, x_2, t), \quad E_2 = E_2(x_1, x_2, t), \quad E_3 = 0,$$
$$H_1 = H_2 = 0, \quad H_3 = H_3(x_1, x_2, t). \tag{7.3.1}$$

Correspondingly, the nontrivial components of S_{ij}, T_{ij}, D_i, and B_i are

$$S_4 = u_{3,2}, \quad S_5 = u_{3,1},$$
$$T_4 = c_{44}u_{3,2} - e_{15}E_2, \quad T_5 = c_{44}u_{3,1} - e_{15}E_1, \tag{7.3.2}$$
$$D_1 = e_{15}u_{3,1} + \varepsilon_{11}E_1, \quad D_2 = e_{15}u_{3,2} + \varepsilon_{11}E_2,$$
$$B_3 = \mu_0 H_3. \tag{7.3.3}$$

The relevant equation of motion and Maxwell's equations take the following form:

$$c_{44}(u_{3,11} + u_{3,22}) - e_{15}(E_{1,1} + E_{2,2}) = \rho\ddot{u}_3,$$
$$e_{15}(u_{3,11} + u_{3,22}) + \varepsilon_{11}(E_{1,1} + E_{2,2}) = 0, \tag{7.3.4}$$

$$E_{2,1} - E_{1,2} = -\mu_0\dot{H}_3,$$
$$H_{3,2} = e_{15}\dot{u}_{3,1} + \varepsilon_{11}\dot{E}_1,$$
$$-H_{3,1} = e_{15}\dot{u}_{3,2} + \varepsilon_{11}\dot{E}_2. \tag{7.3.5}$$

Eliminating the electric field components from Eq. (7.3.4), we obtain

$$\bar{c}_{44}(u_{3,11} + u_{3,22}) = \rho\ddot{u}_3, \tag{7.3.6}$$

where

$$\bar{c}_{44} = c_{44} + e_{15}^2/\varepsilon_{11} = c_{44}(1 + k_{15}^2),$$

$$k_{15}^2 = \frac{e_{15}^2}{\varepsilon_{11}c_{55}} = k^2. \tag{7.3.7}$$

$\bar{c}_{44}$ is a piezoelectrically stiffened shear elastic constant. Differentiating Eq. $(7.3.5)_1$ with respect to time once and substituting from Eqs. $(7.3.5)_{2,3}$, we have

$$H_{3,11} + H_{3,22} = \varepsilon_{11}\mu_0\ddot{H}_3. \tag{7.3.8}$$

The above equations can be written in coordinate-independent forms as

$$v_T^2\nabla^2 u_3 = \ddot{u}_3,$$

$$c^2\nabla^2 H_3 = \ddot{H}_3, \tag{7.3.9}$$

$$\dot{\mathbf{D}} = -\mathbf{e}_3 \times \nabla H_3,$$

where

$$v_T^2 = \frac{\bar{c}_{44}}{\rho}, \quad c^2 = \frac{1}{\varepsilon_{11}\mu_0}. \tag{7.3.10}$$

v_T and c are the speed of plane shear waves and the speed of light in the x_1 direction in the material, respectively. ∇ and ∇^2 are the two-dimensional gradient operator and Laplacian, respectively. $\mathbf{D}$ is the electric displacement vector in the (x_1, x_2) plane. $\mathbf{e}_3$ is the unit vector in the x_3 direction. The magnetic field has one component H_3 only. Therefore, one displacement component and one magnetic field component are sufficient to describe the antiplane mechanical and electromagnetic fields. This is simpler than using the scalar and vector potentials of the electromagnetic fields. In polar coordinates,

Eq. (7.3.9) takes the form of

$$v_T^2 \left(\frac{\partial^2 u_3}{\partial r^2} + \frac{1}{r} \frac{\partial u_3}{\partial r} + \frac{1}{r^2} \frac{\partial^2 u_3}{\partial \theta^2} \right) = \ddot{u}_3, \qquad (7.3.11)$$

$$c^2 \left(\frac{\partial^2 H_3}{\partial r^2} + \frac{1}{r} \frac{\partial H_3}{\partial r} + \frac{1}{r^2} \frac{\partial^2 H_3}{\partial \theta^2} \right) = \ddot{H}_3, \qquad (7.3.12)$$

$$\varepsilon_{11} \dot{E}_r = \frac{1}{r} H_{3,\theta} - e_{15} \dot{u}_{3,r},$$

$$\varepsilon_{11} \dot{E}_\theta = -H_{3,r} - e_{15} \frac{1}{r} \dot{u}_{3,\theta}. \qquad (7.3.13)$$

For later use we also denote

$$\bar{k}_{15}^2 = \frac{e_{15}^2}{\varepsilon_{11} \bar{c}_{55}} = \bar{k}^2 = \frac{k^2}{1 + k^2},$$

$$n^2 = \frac{\varepsilon_{11}}{\varepsilon_0}, \quad c_0^2 = \frac{1}{\varepsilon_0 \mu_0}. \qquad (7.3.14)$$

n is the refractive index of the material. c_0 is the speed of light in a vacuum.

7.4 Surface Waves

Consider a ceramic half-space poled in the x_3 direction (see Fig. 7.1). The surface at $x_2 = 0$ is electroded with a perfect conductor for which we have $E_1 = 0$. The electrode is assumed to be very thin, with negligible mechanical effects such as inertia and stiffness. Hence the surface is traction free with $T_{23} = 0$. We consider antiplane motions near the surface.

We look for surface waves propagating in the x_1 direction [27]. Let

$$u_3 = U \exp(-\xi_2 x_2) \cos(\xi_1 x_1 - \omega t),$$
$$H_3 = H \exp(-\eta_2 x_2) \cos(\xi_1 x_1 - \omega t), \qquad (7.4.1)$$

where U, H, ξ_1, ξ_2, η_2, and ω are undetermined constants. The sub-

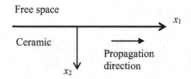

Fig. 7.1. A ceramic half-space.

stitution of Eq. (7.4.1) into Eq. (7.3.9)$_{1,2}$ results in

$$\xi_2^2 = \xi_1^2 - \rho\omega^2/\bar{c}_{44} = \xi_1^2\left(1 - \frac{v^2}{v_T^2}\right) > 0, \qquad (7.4.2)$$

$$\eta_2^2 = \xi_1^2 - \varepsilon_{11}\mu_0\omega^2 = \xi_1^2\left(1 - \frac{v^2}{c^2}\right) > 0. \qquad (7.4.3)$$

v is the surface wave speed. The inequalities are for decaying behaviors from the surface. In this section, we focus on the case of $v < v_T < c$ when both the acoustic and electromagnetic fields or waves are guided by the surface, i.e., they both decay from the surface. The case of $v_T < v < c$, i.e., the electromagnetic wave is guided by the surface but the acoustic wave is not, will be examined later in Section 7.9. The stress and electric field components produced by Eq. (7.4.1) relevant to boundary conditions are

$$T_{23} = -\frac{1}{\varepsilon_{11}\omega}[\varepsilon_{11}\bar{c}_{44}\omega\xi_2 U \exp(-\xi_2 x_2)$$

$$+ e_{15}\xi_1 H \exp(-\eta_2 x_2)]\cos(\xi_1 x_1 - \omega t), \qquad (7.4.4)$$

$$E_1 = \frac{1}{\varepsilon_{11}\omega}[e_{15}\omega\xi_1 U \exp(-\xi_2 x_2)$$

$$+ \eta_2 H \exp(-\eta_2 x_2)] \times \sin(\xi_1 x_1 - \omega t). \qquad (7.4.5)$$

For $E_1 = 0$ and $T_{23} = 0$ at the surface, we have

$$e_{15}\omega\xi_1 U + \eta_2 H = 0,$$

$$\varepsilon_{11}\bar{c}_{44}\omega\xi_2 U + e_{15}\xi_1 H = 0. \qquad (7.4.6)$$

For nontrivial solutions of U and/or H, the determinant of the coefficient matrix of Eq. (7.4.6) has to vanish, which leads to

$$\sqrt{1 - \frac{v^2}{v_T^2}}\sqrt{1 - \frac{v^2}{c^2}} = \bar{k}_{15}^2, \qquad (7.4.7)$$

or

$$\sqrt{1 - \frac{v^2}{v_T^2}}\sqrt{1 - \lambda\frac{v^2}{v_T^2}} = \bar{k}_{15}^2, \qquad (7.4.8)$$

$$\lambda = \frac{v_T^2}{c^2}, \quad c^2 = \frac{1}{\varepsilon_{11}\mu_0}.$$

Equation (7.4.7) is an equation for the surface wave speed v. Waves with their speed determined by Eq. (7.4.7) are clearly nondispersive. Their existence requires a nonzero k_{15}.

For the essentially acoustic wave determined by Eq. (7.4.8), $v \ll c$. Since λ is very small, it is simpler and more revealing to examine the following perturbation solution of Eq. (7.4.7) for small λ:

$$v^2 \cong v_T^2(1 - \bar{k}_{15}^4)(1 - \lambda\bar{k}_{15}^4). \qquad (7.4.9)$$

Equation (7.4.9) shows that the effect of electromagnetic coupling on the acoustic wave speed is on the order of $\lambda\bar{k}_{15}^4$. As a numerical example, we consider polarized ceramics of PZT-7A. Calculation shows that

$$\bar{k}_{15} = 0.671, \quad \lambda\bar{k}_{15}^4 = 6.38 \times 10^{-7}. \qquad (7.4.10)$$

Hence, the modification of the acoustic wave speed due to electromagnetic coupling is very small and is negligible in most applications. When λ is set to zero, or when the speed of light approaches infinity, Eq. (7.4.9) reduces to

$$v^2 \cong v_T^2(1 - \bar{k}_{15}^4), \qquad (7.4.11)$$

which is the speed of the well-known Bleustein–Gulyaev wave [28,29] in quasistatic piezoelectricity.

Piezoelectromagnetic surface waves can also propagate on an unelectroded ceramic half-space [27] with their quasistatic counter parts in [28,29]. These waves will be obtained as a special case of the interface waves in the next section.

7.5 Interface Waves

Antiplane piezoelectromagnetic waves can also propagate near the interface between two ceramic half-spaces (see Fig. 7.2) [30]. The fields decay from the interface exponentially while propagating in the x_1 direction.

In the half-space with $x_2 > 0$, the fields are the same as those in Eqs. (7.4.1)–(7.4.5). For the half-space occupying $x_2 < 0$, we use a superimposed hat to represent the material parameters and undetermined constants. Similar to Eqs. (7.4.1)–(7.4.5), we have

$$u_3 = \hat{U} \exp(\hat{\xi}_2 x_2) \cos(\xi_1 x_1 - \omega t),$$

$$H_3 = \hat{H} \exp(\hat{\eta}_2 x_2) \cos(\xi_1 x_1 - \omega t),$$

(7.5.1)

$$\hat{\xi}_2^2 = \xi_1^2 - \hat{\rho}\omega^2/\hat{\bar{c}}_{44} = \xi_1^2 \left(1 - \frac{v^2}{\hat{v}_T^2}\right) > 0,$$

$$\hat{\eta}_2^2 = \xi_1^2 - \hat{\varepsilon}_{11}\mu_0\omega^2 = \xi_1^2 \left(1 - \frac{v^2}{\hat{c}^2}\right) > 0,$$

(7.5.2)

$$\hat{v}_T^2 = \frac{\hat{\bar{c}}_{44}}{\hat{\rho}}, \quad \hat{c}^2 = \frac{1}{\hat{\varepsilon}_{11}\mu_0},$$

(7.5.3)

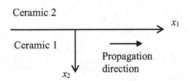

Fig. 7.2.   An interface between two ceramic half-spaces.

$$T_{23} = -\frac{1}{\hat{\varepsilon}_{11}\omega}[-\hat{\varepsilon}_{11}\hat{\bar{c}}_{44}\omega\hat{\xi}_2\hat{U}\exp(\hat{\xi}_2 x_2)$$

$$+ \hat{e}_{15}\xi_1\hat{H}\exp(\hat{\eta}_2 x_2)]\cos(\xi_1 x_1 - \omega t), \qquad (7.5.4)$$

$$E_1 = \frac{1}{\hat{\varepsilon}_{11}\omega}[\hat{e}_{15}\omega\xi_1\hat{U}\exp(\hat{\xi}_2 x_2)$$

$$- \hat{\eta}_2\hat{H}\exp(\hat{\eta}_2 x_2)]\sin(\xi_1 x_1 - \omega t). \qquad (7.5.5)$$

At the interface $x_2 = 0$, the continuity of u_3, H_3, T_{23}, and E_1 need to be imposed. The continuities of u_3 and H_3 imply directly that

$$\hat{U} = U, \quad \hat{H} = H. \qquad (7.5.6)$$

Then the continuity of T_{23} and E_1 results in two linear and homogeneous equations for U and H. For nontrivial solutions the determinant of the coefficient matrix of the equations has to vanish, which leads to the following equation:

$$\left(\bar{c}_{44}\xi_2 + \hat{\bar{c}}_{44}\hat{\xi}_2\right)\left(\frac{\eta_2}{\varepsilon_{11}} + \frac{\hat{\eta}_2}{\hat{\varepsilon}_{11}}\right) = \left(\frac{e_{15}}{\varepsilon_{11}} - \frac{\hat{e}_{15}}{\hat{\varepsilon}_{11}}\right)^2\xi_1^2, \qquad (7.5.7)$$

which is an equation for the wave speed v:

$$\left(\sqrt{1 - \frac{v^2}{v_T^2}} + \frac{\hat{\bar{c}}_{44}}{\bar{c}_{44}}\sqrt{1 - \frac{v^2}{\hat{v}_T^2}}\right)$$

$$\times \left(\sqrt{1 - \frac{v^2}{c^2}} + \frac{\varepsilon_{11}}{\hat{\varepsilon}_{11}}\sqrt{1 - \frac{v^2}{\hat{c}^2}}\right) = \frac{\varepsilon_{11}}{\bar{c}_{44}}\left(\frac{e_{15}}{\varepsilon_{11}} - \frac{\hat{e}_{15}}{\hat{\varepsilon}_{11}}\right)^2. \qquad (7.5.8)$$

Clearly, the wave is nondispersive. In the special case when $x_2 < 0$ is a vacuum, we have

$$\hat{\bar{c}}_{44} = 0, \quad \hat{e}_{15} = 0, \quad \hat{\varepsilon}_{11} = \varepsilon_0. \qquad (7.5.9)$$

In this case Eq. (7.5.8) reduces to the following equation for the piezoelectromagnetic surface wave over an unelectroded ceramic half-space in [27]:

$$\sqrt{1 - \frac{v^2}{v_T^2}} \left(\sqrt{1 - \frac{v^2}{c^2}} + \frac{\varepsilon_{11}}{\varepsilon_0} \sqrt{1 - \frac{v^2}{\hat{c}^2}} \right) = \bar{k}_{15}^2. \qquad (7.5.10)$$

If the two half-spaces are of the same ceramic but are with opposite poling directions, we have

$$\hat{\bar{c}}_{44} = \bar{c}_{44}, \quad \hat{e}_{15} = -e_{15}, \quad \hat{\varepsilon}_{11} = \varepsilon_{11}. \qquad (7.5.11)$$

Then Eq. (7.5.8) simplifies to

$$\sqrt{1 - \frac{v^2}{v_T^2}} \sqrt{1 - \frac{v^2}{c^2}} = \bar{k}_{15}^2. \qquad (7.5.12)$$

Equation (7.5.12) is the same as Eq. (7.4.7) for the piezoelectromagnetic surface wave over an electroded ceramic half-space. In the limit when the speed of light goes to infinity, Eq. (7.5.8) reduces to the following equation for the quasistatic piezoelectric interface wave in [31]

$$\left(\sqrt{1 - \frac{v^2}{v_T^2}} + \frac{\hat{\bar{c}}_{44}}{\bar{c}_{44}} \sqrt{1 - \frac{v^2}{\hat{v}_T^2}} \right) \left(1 + \frac{\varepsilon_{11}}{\hat{\varepsilon}_{11}} \right)$$

$$= \frac{\varepsilon_{11}}{\bar{c}_{44}} \left(\frac{e_{15}}{\varepsilon_{11}} - \frac{\hat{e}_{15}}{\hat{\varepsilon}_{11}} \right)^2. \qquad (7.5.13)$$

7.6 Gap Waves

Since electromagnetic fields can exist in a vacuum, two piezoelectromagnetic half-spaces with a gap in between can interact with each other (see Fig. 7.3). We look for waves propagating near the gap [32].

In the half-space with $x_2 > h$, the fields are the same as those in Eqs. (7.4.1)–(7.4.5). For the half-space occupying $x_2 < -h$, the fields

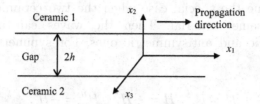

Fig. 7.3. Two ceramic half-spaces with a gap in between.

in Eqs. (7.5.1)–(7.5.5) can still be used. The gap within $-h < x_2 < h$ is a free space with ε_0 and μ_0. The fields in the gap can be written as

$$H_3 = [H' \exp(\eta_2' x_2) + H'' \exp(-\eta_2' x_2)] \cos(\xi_1 x_1 - \omega t), \qquad (7.6.1)$$

where H' and H'' are undetermined constants, and

$$(\eta_2')^2 = \xi_1^2 - \varepsilon_0 \mu_0 \omega^2 = \xi_1^2 \left(1 - \frac{v^2}{c_0^2}\right),$$

$$c_0^2 = \frac{1}{\varepsilon_0 \mu_0}. \qquad (7.6.2)$$

Then, in the gap,

$$E_1 = -\frac{1}{\varepsilon_0 \omega}[\eta_2' H' \exp(\eta_2' x_2)$$

$$- \eta_2' H'' \exp(-\eta_2' x_2)] \sin(\xi_1 x_1 - \omega t). \qquad (7.6.3)$$

At the interfaces at $x_2 = \pm h$, T_{23} vanishes and the continuities of E_1 and H_3 need to be imposed. This leads to six linear and homogeneous equations for U, H, H', H'', $\hat{U}$ and $\hat{H}$. For nontrivial solutions the determinant of the coefficient matrix of the equations has to vanish, which leads to an equation that determines the dispersion relations of the gap waves.

We examine the special case when the two ceramic half-spaces are of the same material. Then the waves can be separated into symmetric and antisymmetric ones. For symmetric waves we consider

$$\hat{U} = U, \quad \hat{H} = -H, \quad H'' = -H'. \tag{7.6.4}$$

The dispersion relation assumes the following simple form:

$$\tanh \xi_1 h \sqrt{1 - \frac{v^2}{c_0^2}} = \frac{n^2 \sqrt{1 - \frac{v^2}{v_T^2}} \sqrt{1 - \frac{v^2}{c_0^2}}}{\bar{k}^2 - \sqrt{1 - \frac{v^2}{v_T^2}} \sqrt{1 - \frac{v^2}{c^2}}}, \tag{7.6.5}$$

where $n^2 = \varepsilon_{11}/\varepsilon_0$. Equation (7.6.5) shows that the wave is dispersive. In the limit when the speed of light goes to infinity, Eq. (7.6.5) reduces to the dispersion relation for the symmetric quasistatic piezoelectric gap wave in [33]. When $\bar{k}$ is set to zero, Eq. (7.6.5) formally reduces to the dispersion relation for guided electromagnetic waves in a plate [34], but such a guided wave exists only when the gap is filled with a material with a dielectric constant larger than that of the ceramic half-space.

For antisymmetric waves we consider

$$\hat{U} = -U, \quad \hat{H} = H, \quad H'' = H'. \tag{7.6.6}$$

In this case we have

$$\tanh \xi_1 h \sqrt{1 - \frac{v^2}{c_0^2}} = \left(\frac{n^2 \sqrt{1 - \frac{v^2}{v_T^2}} \sqrt{1 - \frac{v^2}{c_0^2}}}{\bar{k}^2 - \sqrt{1 - \frac{v^2}{v_T^2}} \sqrt{1 - \frac{v^2}{c^2}}} \right)^{-1}. \tag{7.6.7}$$

Equation (7.6.7) reduces to the dispersion relation for the antisymmetric quasistatic piezoelectric gap wave in [33] when the speed of light goes to infinity. In the special case of $h = 0$, Eq. (7.6.7) reduces to

$$\sqrt{1 - \frac{v^2}{v_T^2}} \sqrt{1 - \frac{v^2}{c^2}} = \bar{k}^2, \tag{7.6.8}$$

which is Eq. (7.4.7) for surface waves over an electroded ceramic half-space. We note that the right-hand side of Eq. (7.6.5) is the inverse of that of Eq. (7.6.7). Therefore, for any value of $v < v_T$, one of the right-hand sides of Eqs. (7.6.5) and (7.6.7) is between minus one and one. Since the range of a hyperbolic tangent function is between minus one and one, a root for ξ_1 can always be found for any value of v from either Eq. (7.6.5) or Eq. (7.6.7).

7.7 Waves in a Plate

A plate may be used as either an acoustic or an optical waveguide. Consider antiplane waves in a plate of polarized ceramics with traction-free surfaces at $x_2 = \pm h$ (Fig. 7.4) [35]. The two surfaces may be both electroded or not, which will be treated separately below. For an electroded plate, the electrodes are shorted and are assumed to be very thin and ideal conductors whose mechanical effects are negligible.

Consider an electroded plate first. For waves propagating in the x_1 direction, we look for

$$u_3 = (U \cos \xi_2 x_2 + V \sin \xi_2 x_2) \cos(\xi_1 x_1 - \omega t),$$
$$H_3 = (G \cosh \eta_2 x_2 + H \sinh \eta_2 x_2) \cos(\xi_1 x_1 - \omega t), \tag{7.7.1}$$

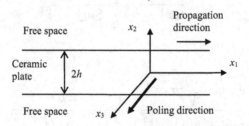

Fig. 7.4. A plate of polarized ceramics.

where

$$\xi_2^2 = \rho\omega^2/\bar{c}_{44} - \xi_1^2 = \xi_1^2\left(\frac{v^2}{v_T^2} - 1\right), \qquad (7.7.2)$$

$$\eta_2^2 = \xi_1^2 - \varepsilon_{11}\mu_0\omega^2 = \xi_1^2\left(1 - \frac{v^2}{c^2}\right). \qquad (7.7.3)$$

We also have

$$T_{23} = \Bigg[-\bar{c}_{44}\xi_2 U \sin\xi_2 x_2 + \bar{c}_{44}\xi_2 V \cos\xi_2 x_2$$

$$- \frac{e_{15}}{\varepsilon_{11}\omega}\xi_1(G\cosh\eta_2 x_2 + H\sinh\eta_2 x_2)\Bigg]\cos(\xi_1 x_1 - \omega t),$$

$$(7.7.4)$$

$$E_1 = \frac{1}{\varepsilon_{11}\omega}(e_{15}\omega\xi_1 U \cos\xi_2 x_2 + e_{15}\omega\xi_1 V \sin\xi_2 x_2$$

$$- G\eta_2\sinh\eta_2 x_2 - H\eta_2\cosh\eta_2 x_2)\sin(\xi_1 x_1 - \omega t). \ (7.7.5)$$

At $x_2 = \pm h$, T_{23} vanishes for thin electrodes and E_1 vanishes for ideal conductors. This leads to four linear and homogeneous equations for U, V, G and H. For nontrivial solutions the determinant of the coefficient matrix of the equations has to vanish, which leads to an equation that determines the dispersion relations. The waves can be separated into symmetric and antisymmetric ones. For symmetric waves we consider the case of $V = 0$ and $G = 0$. The dispersion relation is determined by

$$\frac{\tan\xi_2 h}{\tanh\eta_2 h} = \frac{-\bar{k}^2\xi_1^2}{\xi_2\eta_2}, \qquad (7.7.6)$$

or

$$\frac{\tan\xi_1 h\sqrt{\frac{v^2}{v_T^2} - 1}}{\tanh\xi_1 h\sqrt{1 - \frac{v^2}{c^2}}} = \frac{-\bar{k}^2}{\sqrt{\frac{v^2}{v_T^2} - 1}\sqrt{1 - \frac{v^2}{c^2}}}. \qquad (7.7.7)$$

For antisymmetric waves we consider the case when $U = 0$ and $H = 0$. The dispersion relation is determined by

$$\frac{\tan \xi_2 h}{\tanh \eta_2 h} = \frac{\xi_2 \eta_2}{\bar{k}^2 \xi_1^2},$$ (7.7.8)

or

$$\frac{\tan \xi_1 h \sqrt{\frac{v^2}{v_T^2} - 1}}{\tanh \xi_1 h \sqrt{1 - \frac{v^2}{c^2}}} = \frac{\sqrt{\frac{v^2}{v_T^2} - 1} \sqrt{1 - \frac{v^2}{c^2}}}{\bar{k}^2}.$$ (7.7.9)

Next consider an unelectroded plate. In this case the electromagnetic fields inside the plate are coupled to those in the surrounding free space. For $x_2 < -h$, we have

$$H_3 = H' \exp(\eta_2' x_2) \cos(\xi_1 x_1 - \omega t),$$ (7.7.10)

$$(\eta_2')^2 = \xi_1^2 - \varepsilon_0 \mu_0 \omega^2 = \xi_1^2 \left(1 - \frac{v^2}{c_0^2}\right),$$ (7.7.11)

$$E_1 = -\frac{1}{\varepsilon_0 \omega} \eta_2' H' \exp(\eta_2' x_2) \sin(\xi_1 x_1 - \omega t).$$ (7.7.12)

Similarly, for $x_2 > h$, the fields are

$$H_3 = H'' \exp(-\eta_2' x_2) \cos(\xi_1 x_1 - \omega t),$$ (7.7.13)

$$E_1 = \frac{1}{\varepsilon_0 \omega} \eta_2' H'' \exp(-\eta_2' x_2) \sin(\xi_1 x_1 - \omega t).$$ (7.7.14)

At $x_2 = \pm h$, T_{23} vanishes. The continuities of H_3 and E_1 need to be imposed. The continuity of D_2 is implied by the continuity of H_3. These conditions lead to six linear and homogeneous equations for U, V, G, H, H' and H''. For nontrivial solutions, the determinant of the coefficient matrix of the equations has to vanish. For antisymmetric waves we have $U = 0$, $H = 0$ and $H' = H''$. The dispersion relation is determined by

$$\xi_2(\eta_2 \tanh \eta_2 h + n^2 \eta_2') = \overline{k}^2 \xi_1^2 \tan \xi_2 h, \qquad (7.7.15)$$

or

$$\sqrt{\frac{v^2}{v_T^2} - 1} \left(\sqrt{1 - \frac{v^2}{c^2}} \tanh \xi_1 h \sqrt{1 - \frac{v^2}{c^2}} + n^2 \sqrt{1 - \frac{v^2}{c_0^2}} \right)$$

$$= \overline{k}^2 \tan \xi_1 h \sqrt{\frac{v^2}{v_T^2} - 1}. \qquad (7.7.16)$$

When $\overline{k} = 0$, i.e., the material is not piezoelectric, Eq. (7.7.16) reduces to

$$\sqrt{1 - \frac{v^2}{c^2}} \tanh \xi_1 h \sqrt{1 - \frac{v^2}{c^2}} + n^2 \sqrt{1 - \frac{v^2}{c_0^2}} = 0,$$

$$\text{or} \quad \tan \xi_1 h \sqrt{\frac{v^2}{v_T^2} - 1} = \infty. \qquad (7.7.17)$$

The two equations in Eq. (7.7.17) determine the dispersion relations for electromagnetic waves in a dielectric plate [34] and the dispersion relations for antiplane acoustic waves in an elastic plate (see Section 3.9), respectively. When the speed of light goes to infinity, Eq. (7.7.16) reduces to

$$\sqrt{\frac{v^2}{v_T^2} - 1} \left(\tanh \xi_1 h + n^2 \right) = \overline{k}^2 \tan \xi_1 h \sqrt{\frac{v^2}{v_T^2} - 1}, \qquad (7.7.18)$$

which determines the dispersion relation for quasistatic piezoelectric waves in a ceramic plate [36]. For symmetric waves we consider $V = 0$, $G = 0$ and $H' = -H''$. The dispersion relation is determined by

$$\xi_2(n^2\eta_2' \tanh\eta_2 h + \eta_2)\tan\xi_2 h = -\bar{k}^2\xi_1^2 \tanh\eta_2 h, \qquad (7.7.19)$$

or

$$\sqrt{\frac{v^2}{v_T^2} - 1}\left(n^2\sqrt{1 - \frac{v^2}{c_0^2}}\tanh\xi_1 h\sqrt{1 - \frac{v^2}{c^2}} + \sqrt{1 - \frac{v^2}{c^2}}\right)$$

$$\times \tanh\xi_1 h\sqrt{\frac{v^2}{v_T^2} - 1} = -\bar{k}^2 \tanh\xi_1 h\sqrt{1 - \frac{v^2}{c^2}}. \qquad (7.7.20)$$

We examine Eq. (7.7.19) approximately. Consider long waves with wavelength $\gg$ plate thickness, i.e., $\xi_1 h \ll 1$. We focus on the lowest branch of elastic waves (the so-called face-shear wave). In this case, in terms of the wave speed, v, Eq. (7.7.19) assumes the following form:

$$\left(1 + \xi_1 h n^2\sqrt{1 - \frac{v^2}{c_0^2}}\right)\left(\frac{v^2}{v_T^2} - 1\right) = -\bar{k}^2, \qquad (7.7.21)$$

which may be further approximated by

$$\left(1 + \xi_1 h n^2\right)\left(\frac{v^2}{v_T^2} - 1\right) = -\bar{k}^2. \qquad (7.7.22)$$

Equation (7.7.22) shows that the elastic face-shear wave which is nondispersive becomes dispersive due to piezoelectric and electromagnetic couplings. We plot Eq. (7.7.22) for a few ceramics in Fig. 7.5 where $v_B = v_T$ and $\xi = \xi_1$. The figure is for long waves with a small $\xi_1 h$. $\xi_1 h < 0.1$ translates into $2\pi h/\lambda < 0.1$ or $\lambda/2h > 10\pi$, i.e., the wavelength is about 30 times the plate thickness. The above approximate analysis is not accurate when $\xi_1 h$ is not small.

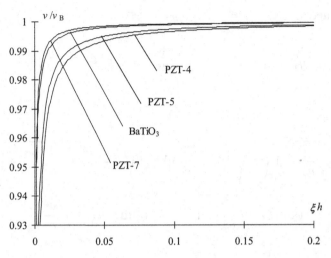

Fig. 7.5. Dispersion of long face-shear waves in an unelectroded plate ($v_B = v_T \cdot \xi = \xi_1$).

7.8 Love Wave

Consider a piezoelectromagnetic half-space carrying an elastic layer (Fig. 7.6) [37]. The elastic layer may be an ideal conductor or a dielectric. The surface at $x_2 = -h$ is traction free. We consider antiplane waves.

First consider the case when the elastic layer is a perfect conductor. We look for waves propagating in the x_1 direction. In the piezoelectric half-space, the possible fields are from Eqs. (7.4.1)–(7.4.5):

$$u_3 = U \exp(-\xi_2 x_2) \cos(\xi_1 x_1 - \omega t), \tag{7.8.1}$$

$$H_3 = H \exp(-\eta_2 x_2) \cos(\xi_1 x_1 - \omega t),$$

$$\xi_2^2 = \xi_1^2 - \rho \omega^2 / \bar{c}_{44} = \xi_1^2 \left(1 - \frac{v^2}{v_T^2} \right) > 0, \tag{7.8.2}$$

$$\eta_2^2 = \xi_1^2 - \varepsilon_{11} \mu_0 \omega^2 = \xi_1^2 \left(1 - \frac{v^2}{c^2} \right) > 0, \tag{7.8.3}$$

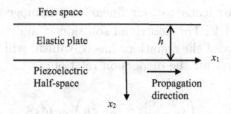

Fig. 7.6. A piezoelectromagnetic half-space with an elastic layer.

$$T_{23} = -\frac{1}{\varepsilon_{11}\omega}[\varepsilon_{11}\bar{c}_{44}\omega\xi_2 U \exp(-\xi_2 x_2)$$

$$+ e_{15}\xi_1 H \exp(-\eta_2 x_2)]\cos(\xi_1 x_1 - \omega t), \quad (7.8.4)$$

$$E_1 = \frac{1}{\varepsilon_{11}\omega}[e_{15}\omega\xi_1 U \exp(-\xi_2 x_2)$$

$$+ \eta_2 H \exp(-\eta_2 x_2)]\sin(\xi_1 x_1 - \omega t). \quad (7.8.5)$$

For the ideal conductor layer, the electric field vanishes everywhere. The elastic fields are

$$u_3 = (\hat{U}\cos\hat{\xi}_2 x_2 + \hat{V}\sin\hat{\xi}_2 x_2)\cos(\xi_1 x_1 - \omega t), \quad (7.8.6)$$

$$\hat{\xi}_2^2 = \hat{\rho}\omega^2/\hat{c}_{44} - \xi_1^2 = \xi_1^2\left(\frac{v^2}{\hat{v}_T^2} - 1\right), \quad (7.8.7)$$

$$\hat{v}_T^2 = \frac{\hat{c}_{44}}{\hat{\rho}}, \quad (7.8.8)$$

$$T_{23} = (-\hat{c}_{44}\hat{\xi}_2\hat{U}\sin\hat{\xi}_2 x_2$$

$$+ \hat{c}_{44}\hat{\xi}_2\hat{V}\cos\hat{\xi}_2 x_2)\cos(\xi_1 x_1 - \omega t). \quad (7.8.9)$$

At the interface $x_2 = 0$ and the boundary $x_2 = -h$, we have the following continuity and boundary conditions:

$$u_3(0^+) = u_3(0^-), \quad T_{23}(0^+) = T_{23}(0^-),$$
$$E_1(0^+) = 0, \quad T_{23}(-h) = 0. \quad (7.8.10)$$

Equation (7.8.10) leads to four linear and homogeneous equations for U, H, $\hat{U}$ and $\hat{V}$. For nontrivial solutions the determinant of the coefficient matrix of the equations has to vanish, which results in the following equation for the dispersion relation:

$$\eta_2 \left(\xi_2 - \frac{\hat{c}_{44}}{\overline{c}_{44}} \hat{\xi}_2 \tan \hat{\xi} h \right) = \overline{k}^2 \xi_1^2, \tag{7.8.11}$$

or

$$\sqrt{1 - \frac{v^2}{c^2}} \left(\sqrt{1 - \frac{v^2}{v_T^2}} \right.$$
$$\left. - \frac{\hat{c}_{44}}{\overline{c}_{44}} \sqrt{\frac{v^2}{\hat{v}_T^2} - 1} \tan \xi_1 h \sqrt{\frac{v^2}{\hat{v}_T^2} - 1} \right) = \overline{k}^2. \tag{7.8.12}$$

The waves are dispersive. When $\overline{k} = 0$, i.e., the material is not piezoelectric, Eq. (7.8.12) reduces to

$$\sqrt{1 - \frac{v^2}{v_T^2}} - \frac{\hat{c}_{44}}{\overline{c}_{44}} \sqrt{\frac{v^2}{\hat{v}_T^2} - 1} \tan \xi_1 h \sqrt{\frac{v^2}{\hat{v}_T^2} - 1} = 0, \tag{7.8.13}$$

which is the well-known equation that determines the speed of Love waves in elasticity (see Eq. (3.10.15)). When the speed of light goes to infinity, Eq. (7.8.12) reduces to

$$\sqrt{1 - \frac{v^2}{v_T^2}} - \frac{\hat{c}_{44}}{\overline{c}_{44}} \sqrt{\frac{v^2}{\hat{v}_T^2} - 1} \tan \xi_1 h \sqrt{\frac{v^2}{\hat{v}_T^2} - 1} = \overline{k}^2, \tag{7.8.14}$$

which determines the dispersion relation for the quasistatic piezoelectric Love waves in a ceramic half-space carrying an elastic metal layer in [38].

Next consider the case when the elastic layer between $-h < x_2 < 0$ is a nonpiezoelectric dielectric and carries a thin and ideal electrode with negligible mass and negligible elastic stiffness at $x_2 = -h$. In addition to the mechanical fields in Eqs. (7.8.6)–(7.8.9), there also exist the following electromagnetic fields in the elastic layer:

$$H_3 = (\hat{G}\cosh\hat{\eta}_2 x_2 + \hat{H}\sinh\hat{\eta}_2 x_2)\cos(\xi_1 x_1 - \omega t), \quad (7.8.15)$$

$$\hat{\eta}_2^2 = \xi_1^2 - \hat{\varepsilon}_{11}\mu_0\omega^2 = \xi_1^2\left(1 - \frac{v^2}{\hat{c}^2}\right), \quad (7.8.16)$$

$$\hat{c}^2 = \frac{1}{\hat{\varepsilon}_{11}\mu_0}, \quad (7.8.17)$$

$$E_1 = \frac{1}{\hat{\varepsilon}_{11}\omega}(-\hat{G}\hat{\eta}_2\sinh\hat{\eta}_2 x_2$$

$$- \hat{H}\hat{\eta}_2\cosh\hat{\eta}_2 x_2)\sin(\xi_1 x_1 - \omega t). \quad (7.8.18)$$

At the interface $x_2 = 0$ and the boundary $x_2 = -h$, we have the following continuity and boundary conditions:

$$u_3(0^+) = u_3(0^-), \quad T_{23}(0^+) = T_{23}(0^-),$$
$$H_3(0^+) = H_3(0^-), \quad E_1(0^+) = E_1(0^-), \quad (7.8.19)$$
$$T_{23}(-h) = 0, \quad E_1(-h) = 0.$$

Equation (7.8.19) leads to six linear and homogeneous equations for U, H, $\hat{U}$, $\hat{V}$, $\hat{G}$ and $\hat{H}$. For nontrivial solutions, the determinant of the coefficient matrix of the equations has to vanish. This yields

$$\left(\eta_2 + \frac{\varepsilon_{11}}{\hat{\varepsilon}_{11}}\hat{\eta}_2\tanh\hat{\eta}_2 h\right)\left(\xi_2 - \frac{\hat{c}_{44}}{\bar{c}_{44}}\hat{\xi}_2\tan\hat{\xi}h\right) = \bar{k}^2\xi_1^2, \quad (7.8.20)$$

or

$$\left(\sqrt{1 - \frac{v^2}{c^2}} + \frac{\varepsilon_{11}}{\hat{\varepsilon}_{11}}\sqrt{1 - \frac{v^2}{\hat{c}^2}}\tanh\xi_1 h\sqrt{1 - \frac{v^2}{\hat{c}^2}}\right)$$

$$\left(\sqrt{1 - \frac{v^2}{v_T^2} - \frac{\hat{c}_{44}}{\bar{c}_{44}}}\sqrt{\frac{v^2}{\hat{v}_T^2} - 1}\tan\xi_1 h\sqrt{\frac{v^2}{\hat{v}_T^2} - 1}\right) = \bar{k}^2. \quad (7.8.21)$$

When $\bar{k} = 0$, i.e., the material of the half-space is nonpiezoelectric, Eq. (7.8.21) reduces to

$$\sqrt{1 - \frac{v^2}{v_T^2} - \frac{\hat{c}_{44}}{\bar{c}_{44}}\sqrt{\frac{v^2}{\hat{v}_T^2} - 1}\tan\xi_1 h\sqrt{\frac{v^2}{\hat{v}_T^2} - 1}} = 0,$$

$$\text{or} \quad \sqrt{1 - \frac{v^2}{c^2}} + \frac{\varepsilon_{11}}{\hat{\varepsilon}_{11}}\sqrt{1 - \frac{v^2}{\hat{c}^2}}\tanh\xi_1 h\sqrt{1 - \frac{v^2}{\hat{c}^2}} = 0,$$

(7.8.22)

which determine the speed of Love waves in elasticity and the speed of electromagnetic waves in a dielectric layer on a half-space. When the speed of light goes to infinity, Eq. (7.8.21) reduces to

$$\left(1 + \frac{\varepsilon_{11}}{\hat{\varepsilon}_{11}}\tanh\xi_1 h\right)\left(\sqrt{1 - \frac{v^2}{v_T^2}}\right.$$

$$\left. - \frac{\hat{c}_{44}}{\bar{c}_{44}}\sqrt{\frac{v^2}{\hat{v}_T^2} - 1}\tan\xi_1 h\sqrt{\frac{v^2}{\hat{v}_T^2} - 1}\right) = \bar{k}^2,$$

(7.8.23)

which determines the dispersion relation for quasistatic piezoelectric Love waves in a ceramic half-space carrying an elastic dielectric layer.

7.9 Acoustic Leakage in Optical Waveguides

Antiplane piezoelectromagnetic waves in a ceramic plate between two ceramic half-spaces (see Fig. 7.7) have also been studied [39]. Some of the results in the previous sections of this chapter are special cases of the waves in [39]. The structure in Fig. 7.7 can be viewed as an acoustic or electromagnetic (optical) waveguide. In this section, we use the result of [39] to study the structure as an optical waveguide [40].

Designs of electromagnetic waveguides and resonators are routinely performed using Maxwell's equations only without acousto-optic interactions [34]. If the materials of the devices have piezoelectric couplings and the devices are mounted on elastic substrates of

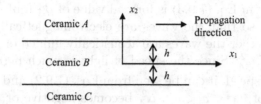

Fig. 7.7. A ceramic plate between two ceramic half-spaces.

other materials, the operating electromagnetic waves may be accompanied by acoustic waves interacting with the substrates. Hence there is a possibility of acoustic leakage or radiation of acoustic energy from an electromagnetic waveguide. Consider a special case of the structure in Fig. 7.7 with ceramics A and C being the same. For symmetric waves propagating in the x_1 direction with a factor of $\cos(\xi x_1 - \omega t)$, the corresponding dispersion relations are determined by [39]:

$$\left(\frac{\beta_A}{\varepsilon_A} \tanh \beta_B h + \frac{\beta_B}{\varepsilon_B}\right)(\bar{c}_A \alpha_A - \bar{c}_B \alpha_B \tan \alpha_B h)$$

$$= \left(\frac{e_A}{\varepsilon_A} - \frac{e_B}{\varepsilon_B}\right)^2 \xi^2 \tanh \beta_B h, \tag{7.9.1}$$

where

$$\alpha_A^2 = \xi^2 - \rho_A \omega^2 / \bar{c}_A = \xi^2 \left(1 - \frac{v^2}{v_A^2}\right), \tag{7.9.2}$$

$$\beta_A^2 = \xi^2 - \varepsilon_A \mu_0 \omega^2 = \xi^2 \left(1 - \frac{v^2}{c_A^2}\right) > 0, \tag{7.9.3}$$

$$\alpha_B^2 = \rho_B \omega^2 / \bar{c}_B - \xi^2 = \xi^2 \left(\frac{v^2}{v_B^2} - 1\right), \tag{7.9.4}$$

$$\beta_B^2 = \xi^2 - \varepsilon_B \mu_0 \omega^2 = \xi^2 \left(1 - \frac{v^2}{c_B^2}\right). \tag{7.9.5}$$

The inequality in Eq. (7.9.3) is for real value of β_A (only the positive root is taken) so that the waves are electromagnetically guided by the plate. Whether the waves are acoustically guided or not depends on the sign of α_A^2. Since the speed of light is much higher than the acoustic wave speed, it can be seen from Eqs. (7.9.2) and (7.9.3) that, in the range of $v_A < v < c_A$, α_A^2 becomes negative or α_A becomes purely imaginary while β_A is still real and positive. In this case, the electromagnetic waves are still guided; but the acoustic fields are not. Consider the case when $v_A < v < c_A$. To be specific, we limit ourselves to the case when $v > c_B$. Together with $v_A < v < c_A$, we are considering v in the range of

$$v_A < c_B < v < c_A. \tag{7.9.6}$$

For convenience we denote

$$\alpha_A^2 = \xi^2 \left(1 - \frac{v^2}{v_A^2} \right) = -\hat{\alpha}_A^2, \quad \alpha_A = i\hat{\alpha}_A,$$

$$\beta_B^2 = \xi^2 \left(1 - \frac{v^2}{c_B^2} \right) = -\hat{\beta}_B^2, \quad \beta_B = i\hat{\beta}_B. \tag{7.9.7}$$

With Eq. (7.9.7), we can write Eq. (7.9.1) as

$$\frac{\beta_A}{\varepsilon_A} \tan \hat{\beta}_B h + \frac{\hat{\beta}_B}{\varepsilon_B}$$

$$= \left(\frac{e_A}{\varepsilon_A} - \frac{e_B}{\varepsilon_B} \right)^2 \frac{\xi^2 \tan \hat{\beta}_B h}{\bar{c}_A i \hat{\alpha}_A - \bar{c}_B \alpha_B \tan \alpha_B h}, \tag{7.9.8}$$

where we have used the identity that $\tanh(iZ) = i\tan(Z)$ for a complex variable Z. The right-hand side of Eq. (7.9.8) is due to piezo-electric coupling which may be a small effect. We use an iteration procedure to solve Eq. (7.9.8). As the lowest order of approximation, we neglect the right-hand side of Eq. (7.9.8) and denote the electromagnetic wave frequencies determined from the left-hand side of

Eq. (7.9.8) by ω_0:

$$\frac{\beta_{A0}}{\varepsilon_A} \tan \hat{\beta}_{B0} h + \frac{\hat{\beta}_{B0}}{\varepsilon_B} \cong 0, \tag{7.9.9}$$

where

$$\beta_{A0} = \sqrt{\xi^2 - \frac{\omega_0^2}{c_A^2}}, \quad \hat{\beta}_{B0} = \sqrt{\frac{\omega_0^2}{c_B^2} - \xi^2}. \tag{7.9.10}$$

Given a wave number ξ, Eq. (7.9.9) determines a series of frequencies $\omega_0(\xi)$ for guided electromagnetic waves. When piezoelectric coupling is considered, these electromagnetic wave frequencies are perturbed a little and are determined by Eq. (7.9.8). Let the frequencies from Eq. (7.9.8) be denoted by

$$\omega = \omega_0 + \Delta\omega. \tag{7.9.11}$$

Substituting Eq. (7.9.11) into Eq. (7.9.8), we obtain the following first-order modification of the electromagnetic wave frequencies due to piezoelectric coupling:

$$\frac{\Delta\omega}{\omega_0} \cong \frac{1}{\omega_0^2} \left(\frac{e_A}{\varepsilon_A} - \frac{e_B}{\varepsilon_B} \right)^2$$

$$\times \frac{\xi^2 \tan(\hat{\beta}_{B0} h)[-\bar{c}_B \alpha_{B0} \tan(\alpha_{B0} h) - \bar{c}_A i \hat{\alpha}_{A0}]}{(\bar{c}_A \hat{\alpha}_{A0})^2 + [\bar{c}_B \alpha_{B0} \tan(\alpha_{B0} h)]^2}$$

$$\times \left(\frac{\beta_{A0} h}{\varepsilon_A \hat{\beta}_{B0} c_B^2 \cos^2(\hat{\beta}_{B0} h)} - \frac{\tan(\hat{\beta}_{B0} h)}{\varepsilon_A \beta_{A0} c_A^2} + \frac{1}{\varepsilon_B \hat{\beta}_{B0} c_B^2} \right)^{-1}, \tag{7.9.12}$$

where

$$\hat{\alpha}_{A0} = \sqrt{\frac{\omega_0^2}{v_A^2} - \xi^2}, \quad \alpha_{B0} = \sqrt{\frac{\omega_0^2}{v_B^2} - \xi^2}. \tag{7.9.13}$$

Equation (7.9.12) gives the frequency perturbation of the electromagnetic waves due to piezoelectric coupling. It is a complex number. In

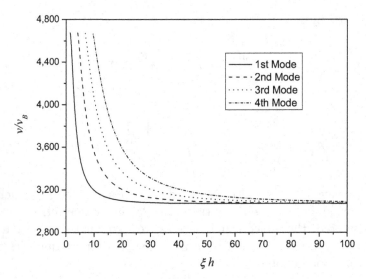

Fig. 7.8.   Dispersion relations of guided electromagnetic waves from Eq. (7.9.9).

addition to its real part representing a frequency shift or additional dispersion, its imaginary part describes damped waves due to acoustic radiation. As a numerical example, consider PZT-4 and PZT-5H for materials A and B which satisfy Eq. (7.9.6).

The first few branches of the dispersion relation for guided electromagnetic waves, i.e., solutions to Eq. (7.9.9), are shown in Fig. 7.8. The way we order the waves or modes is such that higher-order modes have higher frequencies. These waves are dispersive, especially long waves with small wave numbers. The range of the wave speed is bounded by Eq. (7.9.6) from below and above.

When the waves represented by the dispersion relations in Fig. 7.8 are substituted into the right-hand side of Eq. (7.9.12), the imaginary part of the left-hand side of Eq. (7.9.12), which represents dissipation, is shown in Fig. 7.9. The curves in Fig. 7.9 are always positive, indicating damped waves rather than growing waves. The relative decay is on the order of 10^{-6}.

Figure 7.9 is for a very small range of wave speed and the curves already have quite a few oscillations. Curves corresponding to the higher-order modes with higher frequencies have more oscillations. We are interested in the behavior of these curves over the entire range as bounded by Eq. (7.9.6). In Fig. 7.10, we plot the peak values of

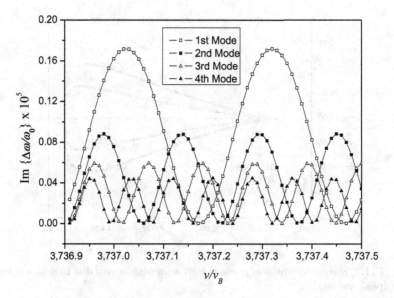

Fig. 7.9. Dissipation of guided electromagnetic waves due to acoustic leakage.

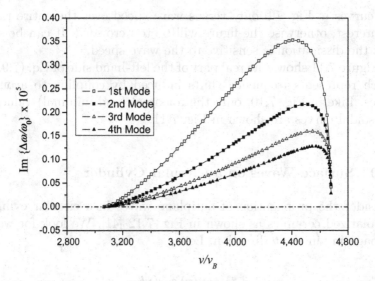

Fig. 7.10. Dissipation of guided electromagnetic waves due to acoustic leakage (peak values).

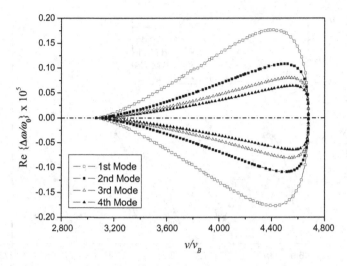

Fig. 7.11.   Frequency shifts of guided electromagnetic waves due to acoustic leakage (peak values).

the curves in Fig. 7.9 only versus wave speed over the entire range of interest, otherwise the figure will be too crowded. It can be seen that the dissipation is sensitive to the wave speed.

Figure 7.11 shows the real part of the left-hand side of Eq. (7.9.12) which represents frequency shifts induced by coupling to acoustic waves. Like in Fig. 7.10, only the maximal (and minimal) values of sinusoidal curves are shown in Fig. 7.11.

7.10   Surface Waves on a Circular Cylinder

Consider the propagation of antiplane waves in a circular cylinder of polarized ceramics as shown in Fig. 7.12 [41]. We look for waves propagating in the θ direction. Let

$$u_3(r,\theta,t) = u(r)\cos(\nu\theta - \omega t),$$
$$H_3(r,\theta,t) = H(r)\cos(\nu\theta - \omega t),$$

(7.10.1)

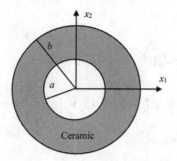

Fig. 7.12. A circular ceramic cylinder.

where ν is a positive integer. The substitution of Eq. (7.10.1) into Eqs. (7.3.11) and (7.3.12) results in

$$\frac{\partial^2 u}{\partial r^2} + \frac{1}{r}\frac{\partial u}{\partial r} + (\alpha^2 - \frac{\nu^2}{r^2})u = 0,$$
$$\frac{\partial^2 H}{\partial r^2} + \frac{1}{r}\frac{\partial H}{\partial r} + (\beta^2 - \frac{\nu^2}{r^2})H = 0,$$

(7.10.2)

where we have denoted

$$\alpha = \frac{\omega}{v_T}, \quad \beta = \frac{\omega}{c}.$$

(7.10.3)

Equation (7.10.2) can be written as Bessel's equations of order ν. Then the general solutions for u_3 and H_3 can be written as

$$u_3 = [C_1 J_\nu(\alpha r) + C_2 Y_\nu(\alpha r)]\cos(\nu\theta - \omega t),$$
$$H_3 = [C_3 J_\nu(\beta r) + C_4 Y_\nu(\beta r)]\cos(\nu\theta - \omega t),$$

(7.10.4)

where J_ν and Y_ν are the νth order Bessel functions of the first and second kinds. C_1 through C_4 are undetermined constants. From Eq. (7.10.4) we obtain the following expressions needed for boundary conditions:

$$E_\theta = \left\{ \frac{\beta}{\varepsilon_{11}\omega}[C_3 J_\nu'(\beta r) + C_4 Y_\nu'(\beta r)] \right.$$

$$\left. + \frac{e_{15}}{\varepsilon_{11}}\frac{\nu}{r}[C_1 J_\nu(\alpha r) + C_2 Y_\nu(\alpha r)] \right\} \sin(\nu\theta - \omega t), \quad (7.10.5)$$

$$T_{rz} = c_{44}u_{3,r} - e_{15}E_r$$

$$= \left\{ \bar{c}_{44}\alpha[C_1 J_\nu'(\alpha r) + C_2 Y_\nu'(\alpha r)] \right.$$

$$\left. + \frac{e_{15}\nu}{\varepsilon_{11}\omega}\frac{1}{r}[C_3 J_\nu(\beta r) + C_4 Y_\nu(\beta r)] \right\} \cos(\nu\theta - \omega t),$$

$$(7.10.6)$$

where a prime indicates differentiation with respect to the whole argument of a function. Consider a solid cylinder ($a = 0$ in Fig. 7.12). Since Y_ν is singular at the origin, terms associated with C_2 and C_4 have to be dropped. Furthermore,

$$J_\nu(x) \to \frac{x^\nu}{2^\nu \Gamma(1+\nu)}, \quad x \to 0. \quad (7.10.7)$$

Let the surface at $r = b$ be traction free and carries a very thin electrode of a perfect conductor. Then T_{rz} and E_θ both vanish on the surface. This leads to the following linear and homogeneous equations for C_1 and C_3:

$$\frac{\beta}{\varepsilon_{11}\omega}C_3 J_\nu'(\beta b) + \frac{e_{15}}{\varepsilon_{11}}\frac{\nu}{b}C_1 J_\nu(\alpha b) = 0, \quad (7.10.8)$$

$$\bar{c}_{44}\alpha C_1 J_\nu'(\alpha b) + \frac{e_{15}\nu}{\varepsilon_{11}\omega}\frac{1}{b}C_3 J_\nu(\beta b) = 0. \quad (7.10.9)$$

For nontrivial solutions, the determinant of the coefficient matrix of the equations has to vanish, which leads to the following equation that determines the wave speed:

$$\frac{ab\beta b J_\nu'(\alpha b) J_\nu'(\beta b)}{\nu^2 J_\nu(\alpha b) J_\nu(\beta b)} = \bar{k}_{15}^2. \qquad (7.10.10)$$

When the speed of light c goes to infinity, β goes to zero. In this case, with the asymptotic expression in Eq. (7.10.7), we can write Eq. (7.10.10) as

$$\frac{ab J_\nu'(\alpha b)}{\nu J_\nu(\alpha b)} = \bar{k}_{15}^2, \qquad (7.10.11)$$

which is the dispersion relation for quasistatic piezoelectric surface waves over a ceramic cylinder [42]. When the wavelength is much smaller the radius of the cylinder b, the cylinder is effectively like a half-space. In this case it is helpful to introduce a surface wave number ξ and a surface wave speed V by

$$\xi = \frac{\nu}{b}, \quad V = \frac{\omega}{\xi}. \qquad (7.10.12)$$

Consider the limit when $\nu \to \infty$ and $b \to \infty$ but ξ remains finite. We have

$$\alpha b = \frac{\omega}{v_T} b = \frac{\omega}{v_T} \frac{b}{\nu} \nu = \frac{\omega}{v_T} \frac{1}{\xi} \nu = \frac{V}{v_T} \nu,$$

$$\beta b = \frac{V}{c} \nu. \qquad (7.10.13)$$

With the following asymptotic expression for Bessel functions of large orders:

$$J_\nu(\nu x) \to \frac{x^\nu \exp(\nu\sqrt{1-x^2})}{\sqrt{2\pi\nu} \sqrt[4]{1-x^2}(1+\sqrt{1-x^2})^\nu}, \qquad (7.10.14)$$

$$\nu \to \infty, \quad 0 < x < 1,$$

Eq. (7.10.10) reduces to

$$\sqrt{1 - \frac{V^2}{v_T^2}} \sqrt{1 - \frac{V^2}{c^2}} = \bar{k}_{15}^2, \qquad (7.10.15)$$

which is Eq. (7.4.7) for the speed of piezoelectromagnetic surface waves in a ceramic half-space.

7.11   Electromagnetic Radiation

A vibrating piezoelectric body radiates electromagnetic waves and power. Thus a vibrating piezoelectric crystal can be used as an electromagnetic antenna whose properties can be designed by the crystal material and geometry. For the same frequency, a vibrating piezoelectric crystal as an acoustic wave resonator is much smaller than an electromagnetic wave resonator. Therefore, vibrating piezoelectric crystals may be suitable for small and low-frequency antennae. The electromagnetic radiation from a vibrating piezoelectric acoustic wave resonator is out of the theory of quasistatic piezoelectricity. However, the theory of piezoelectromagnetism is sufficient for describing the phenomenon. Radiated electromagnetic wave and power from unbounded vibrating crystal plates were analyzed in [24,25]. For a finite vibrating piezoelectric plate, the radiated power was calculated in [43] using the electric dipole approximation of a vibrating crystal. In this section we consider electromagnetic radiation from a vibrating circular cylinder of ceramics poled in the x_3 direction as shown in Fig. 7.13. The cylinder is mechanically driven at $r = b$ by a time-harmonic shear stress, τ. The surface at $r = b$ is unelectroded. Electromagnetic waves propagate away from the cylinder.

For the special case of a solid cylinder $(a = 0)$, the governing equations are

$$v_T^2 \nabla^2 u_3 = \ddot{u}_3, \quad c^2 \nabla^2 H_3 = \ddot{H}_3, \quad r < a,$$

$$c_0^2 \nabla^2 H_3 = \ddot{H}_3, \quad r > a. \tag{7.11.1}$$

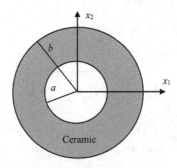

Fig. 7.13.   A circular ceramics cylinder with axial poling.

The boundary and continuity conditions are

$$u_3, H_3 \quad \text{are finite}, \quad r = 0,$$
$$H_3 \quad \text{is outgoing}, \quad r \to \infty,$$

(7.11.2)

$$T_{r3} = \tau \sin \nu\theta \exp(-i\omega t), \quad r = b,$$
$$H_3, E_\theta \quad \text{are continuous}, \quad r = b,$$

(7.11.3)

where ν is a positive integer. For the fields inside the cylinder, we have

$$v_T^2 \left(\frac{\partial^2 u_3}{\partial r^2} + \frac{1}{r} \frac{\partial u_3}{\partial r} + \frac{1}{r^2} \frac{\partial^2 u_3}{\partial \theta^2} \right) = \ddot{u}_3,$$

$$c^2 \left(\frac{\partial^2 H_3}{\partial r^2} + \frac{1}{r} \frac{\partial H_3}{\partial r} + \frac{1}{r^2} \frac{\partial^2 H_3}{\partial \theta^2} \right) = \ddot{H}_3,$$

(7.11.4)

and

$$\varepsilon_{11} \dot{E}_r = \frac{1}{r} H_{3,\theta} - e_{15} \dot{u}_{3,r},$$

$$\varepsilon_{11} \dot{E}_\theta = -H_{3,r} - e_{15} \frac{1}{r} \dot{u}_{3,\theta}.$$

(7.11.5)

Consider

$$u_3(r, \theta, t) = u(r) \sin \nu\theta \exp(-i\omega t),$$

$$H_3(r, \theta, t) = H(r) \cos \nu\theta \exp(-i\omega t).$$

(7.11.6)

The substitution of Eq. (7.11.6) into Eq. (7.11.4) results in

$$\frac{\partial^2 u}{\partial r^2} + \frac{1}{r} \frac{\partial u}{\partial r} + \left(\alpha^2 - \frac{\nu^2}{r^2} \right) u = 0,$$

$$\frac{\partial^2 H}{\partial r^2} + \frac{1}{r} \frac{\partial H}{\partial r} + \left(\beta^2 - \frac{\nu^2}{r^2} \right) H = 0,$$

(7.11.7)

where we have denoted

$$\alpha = \frac{\omega}{v_T}, \quad \beta = \frac{\omega}{c}.$$

(7.11.8)

Equation (7.11.7) can be written as Bessel's equations of order ν. Then the general solutions for u_3 and H_3 can be written as

$$
\begin{aligned}
u_3 &= [C_1 J_\nu(\alpha r) + C_2 Y_\nu(\alpha r)] \sin \nu\theta \exp(-i\omega t), \\
H_3 &= [C_3 J_\nu(\beta r) + C_4 Y_\nu(\beta r)] \cos \nu\theta \exp(-i\omega t),
\end{aligned}
\tag{7.11.9}
$$

where J_ν and Y_ν are the νth order Bessel functions of the first and second kinds. C_1 through C_4 are undetermined constants. Since Y_ν is singular at the origin, terms associated with C_2 and C_4 have to be dropped. Similarly, in the free space of $r > b$, the electromagnetic fields are represented by

$$
H_3 = [C_5 H_\nu^{(1)}(\gamma r) + C_6 H_\nu^{(2)}(\gamma r)] \cos \nu\theta \exp(-i\omega t), \tag{7.11.10}
$$

where $H_\nu^{(1)}$ and $H_\nu^{(2)}$ are the νth order Hankel functions of the first and second kinds, and

$$
\gamma = \frac{\omega}{c_0}. \tag{7.11.11}
$$

To satisfy the radiation condition for outgoing waves at $r \to \infty$, we must have $C_6 = 0$. From the boundary and continuity conditions at $r = b$, we determine

$$
C_1 = \frac{\beta b J_\nu'(\beta b) H_\nu^{(1)}(\gamma b) - \frac{\varepsilon_{11}}{\varepsilon_0} \gamma b J_\nu(\beta b) H_\nu'^{(1)}(\gamma b)}{\Delta} \frac{\tau b}{\bar{c}_{44}},
\tag{7.11.12}
$$

$$
C_3 = i\omega e_{15} b \frac{\nu J_\nu(\alpha b) H_\nu^{(1)}(\gamma b)}{\Delta} \frac{\tau}{\bar{c}_{44}},
\tag{7.11.13}
$$

$$
C_5 = i\omega e_{15} b \frac{\nu J_\nu(\alpha b) J_\nu(\beta b)}{\Delta} \frac{\tau}{\bar{c}_{44}},
\tag{7.11.14}
$$

where

$$\Delta = \alpha b \beta b J_\nu'(\alpha b) J_\nu'(\beta b) H_\nu^{(1)}(\gamma b)$$

$$- \bar{k}_{15}^2 \nu^2 J_\nu(\alpha b) J_\nu(\beta b) H_\nu^{(1)}(\gamma b)$$

$$- \frac{\varepsilon_{11}}{\varepsilon_0} \alpha b \gamma b J_\nu'(\alpha b) J_\nu(\beta b) H_\nu'^{(1)}(\gamma b). \tag{7.11.15}$$

$\Delta = 0$ yields the frequency equation that determines resonances.

We calculate the radiation at the far field with large r using the following asymptotic expressions of Hankel functions with large arguments:

$$H_v^{(1)}(x) \cong \sqrt{\frac{2}{\pi x}} \exp i \left(x - \frac{\nu\pi}{2} - \frac{\pi}{4} \right),$$

$$H_v'^{(1)}(x) \cong i\sqrt{\frac{2}{\pi x}} \exp i \left(x - \frac{\nu\pi}{2} - \frac{\pi}{4} \right). \tag{7.11.16}$$

Then

$$H_3 \cong C_5 \sqrt{\frac{2}{\pi\gamma r}} \exp i \left(\gamma r - \frac{\nu\pi}{2} - \frac{\pi}{4} \right) \tag{7.11.17}$$

$$\times \cos \nu\theta \exp(-i\omega t),$$

$$E_\theta \cong \frac{\gamma}{\omega\varepsilon_0} C_5 \sqrt{\frac{2}{\pi\gamma r}} \exp i \left(\gamma r - \frac{\nu\pi}{2} - \frac{\pi}{4} \right) \tag{7.11.18}$$

$$\times \cos \nu\theta \exp(-i\omega t),$$

which are outgoing. To calculate the radiated power, we need the radial component of the Poynting vector which, with the complex notation, when averaged over a period, is given by

$$S_r = \frac{1}{2}(\mathbf{E}^* \times \mathbf{H})_r = \frac{1}{2}\mathrm{Re}\{E_\theta^* H_3\} = \frac{C_5 C_5^*}{\pi\omega\varepsilon_0 r} \cos^2 \nu\theta, \tag{7.11.19}$$

where an asterisk represents complex conjugate. Equation (7.11.19) shows that the energy flux is inversely proportional to r. It also shows the angular distribution of the radiation. The radiated power per unit length of the cylinder is

$$S = \int_0^{2\pi} S_r r d\theta = \frac{C_5 C_5^*}{\omega \varepsilon_0}. \qquad (7.11.20)$$

We are interested in the frequency range of acoustic waves. Therefore, αb is finite, $\beta b \ll 1$, and $\gamma b \ll 1$. For small arguments, we have

$$J_\nu(x) \cong \frac{x^\nu}{2^\nu \Gamma(1+\nu)}, \quad H_\nu^{(1)}(x) \cong -i \frac{2^\nu \Gamma(\nu)}{\pi x^\nu},$$

$$\frac{x J_\nu'(x)}{J_\nu(x)} \cong \nu, \quad \frac{x H_\nu'^{(1)}(x)}{H_\nu^{(1)}(x)} \cong -\nu. \qquad (7.11.21)$$

Then, approximately,

$$C_5 = \frac{i\omega e_{15} b J_\nu(\alpha b)}{(1 + \frac{\varepsilon_{11}}{\varepsilon_0})\alpha b J_\nu'(\alpha b) - \overline{k}_{15}^2 \nu J_\nu(\alpha b)} \frac{\tau}{\overline{c}_{44}} \frac{1}{H_\nu^{(1)}(\gamma b)}. \qquad (7.11.22)$$

The denominator of the first factor of the right-hand side of Eq. (7.11.22) represents the frequency equation for quasistatic piezo-electric resonances of the cylinder. With Eq. (7.11.22), the radiated power can be written as

$$S = \frac{\omega e_{15}^2 b^2}{2\pi \varepsilon_0} \left| \frac{J_\nu(\alpha b)}{(1 + \frac{\varepsilon_{11}}{\varepsilon_0})\alpha b J_\nu'(\alpha b) - \overline{k}_{15}^2 \nu J_\nu(\alpha b)} \frac{\tau}{\overline{c}_{44}} \right|^2$$

$$\times \frac{2\pi}{H_\nu^{(1)}(\gamma b)[H_\nu^{(1)}(\gamma b)]^*}. \qquad (7.11.23)$$

Equation (7.11.23) shows that S is formally proportional to the square of the piezoelectric constant. It can also be seen that S is large near resonance frequencies. Solutions to more piezoelectromagnetic problems can be found in [44].

7.12 Quasistatic Approximation

The quasistatic approximation in the theory of piezoelectricity can be considered as the lowest-order approximation of a perturbation procedure based on the fact that the acoustic wave speed is much smaller than the speed of light [19]. To see this, we consider a nonmagnetizable piezoelectromagnetic body in a source-free region. First, we write the relevant equations of piezoelectromagnetism into the following three equations for $\mathbf{u}$, $\mathbf{E}$, and $\mathbf{B}$:

$$c_{ijkl}u_{k,li} - e_{kij}E_{k,i} = \rho\ddot{u}_j,$$

$$\varepsilon_{ijk}E_{k,j} = -\dot{B}_i,$$

$$\frac{1}{\mu_0}\varepsilon_{ijk}B_{k,j} = e_{ikl}\dot{u}_{k,l} + \varepsilon_{ik}\dot{E}_k. \tag{7.12.1}$$

Consider an acoustic wave with frequency ω in a piezoelectric crystal of size L. We scale the various independent and dependent variables with respect to characteristic quantities

$$\xi_i = \frac{x_i}{L}, \quad \tau = \omega t,$$

$$U_i = \frac{u_i}{L}, \quad b_i = cB_i, \tag{7.12.2}$$

where

$$c = \frac{1}{\sqrt{\varepsilon_0\mu_0}} \tag{7.12.3}$$

is the speed of light in free space. Then Eq. (7.12.1) takes the following form:

$$\frac{1}{L}c_{ijkl}\frac{\partial^2 U_k}{\partial\xi_l\partial\xi_i} - \frac{1}{L}e_{kij}\frac{\partial E_k}{\partial\xi_i} = \rho\omega^2 L\frac{\partial^2 U_j}{\partial\tau^2}, \tag{7.12.4}$$

$$\frac{1}{L}\varepsilon_{ijk}\frac{\partial E_k}{\partial\xi_j} = -\frac{\omega}{c}\frac{\partial b_i}{\partial\tau}, \tag{7.12.5}$$

$$\frac{1}{cL\mu_0}\varepsilon_{ijk}\frac{\partial b_k}{\partial\xi_j} = \omega\varepsilon_0\frac{e_{ikl}}{\varepsilon_0}\frac{\partial^2 U_k}{\partial\xi_l\partial\tau} + \omega\varepsilon_0\frac{\varepsilon_{ik}}{\varepsilon_0}\frac{\partial E_k}{\partial\tau}, \tag{7.12.6}$$

or

$$c_{ijkl}\frac{\partial^2 U_k}{\partial\xi_l\partial\xi_i} - e_{kij}\frac{\partial E_k}{\partial\xi_i} = \rho\omega^2 L^2\frac{\partial^2 U_j}{\partial\tau^2},\qquad (7.12.7)$$

$$\varepsilon_{ijk}\frac{\partial E_k}{\partial\xi_j} = -\eta\frac{\partial b_i}{\partial\tau},\qquad (7.12.8)$$

$$\varepsilon_{ijk}\frac{\partial b_k}{\partial\xi_j} = \eta\left(\frac{e_{ikl}}{\varepsilon_0}\frac{\partial^2 U_k}{\partial\xi_l\partial\tau} + \frac{\varepsilon_{ik}}{\varepsilon_0}\frac{\partial E_k}{\partial\tau}\right),\qquad (7.12.9)$$

where

$$\eta = \frac{\omega L}{c} \ll 1.\qquad (7.12.10)$$

To the lowest order,

$$c_{ijkl}\frac{\partial^2 U_k}{\partial\xi_l\partial\xi_i} - e_{kij}\frac{\partial E_k}{\partial\xi_i} = \rho\omega^2 L^2\frac{\partial^2 U_j}{\partial\tau^2},\qquad (7.12.11)$$

$$\varepsilon_{ijk}\frac{\partial E_k}{\partial\xi_j} = 0,\qquad (7.12.12)$$

$$\varepsilon_{ijk}\frac{\partial b_k}{\partial\xi_j} = 0,\qquad (7.12.13)$$

or

$$c_{ijkl}u_{k,li} - e_{kij}E_{k,i} = \rho\ddot{u}_j,$$
$$\varepsilon_{ijk}E_{k,j} = 0,\qquad (7.12.14)$$
$$\varepsilon_{ijk}H_{k,j} = 0,$$

which describes quasistatic piezoelectricity.

Chapter 8

Thermal and Dissipative Effects

In this chapter, we generalize the theoretical framework of nonlinear electromagnetoelasticity developed in Chapters 4–6 to include thermal and dissipative effects such as heat conduction and viscosity [18,20]. Electrical conduction is beyond the two-continuum model in Chapter 4 and is treated in the last section of this chapter where a three-continuum model [45] is introduced, with the third continuum being a free charge continuum for describing electrical conduction [45].

8.1 Integral Balance Laws

The integral balance laws in Eqs. (5.1.1)–(5.1.4), (5.1.11), (5.1.13) and (5.1.19) as well as the relationships in Eq. (5.1.5) remain the same:

$$\oint_c \mathbf{E} \cdot d\mathbf{y} = -\frac{\partial}{\partial t} \int_s \mathbf{n} \cdot \mathbf{B} ds, \qquad (8.1.1)$$

$$\oint_c \mathbf{H} \cdot d\mathbf{y} = \frac{\partial}{\partial t} \int_s \mathbf{n} \cdot \mathbf{D} ds, \qquad (8.1.2)$$

$$\int_s \mathbf{n} \cdot \mathbf{D} ds = 0, \qquad (8.1.3)$$

$$\int_s \mathbf{n} \cdot \mathbf{B} ds = 0, \qquad (8.1.4)$$

$$\mathbf{D} = \varepsilon_0 \mathbf{E} + \mathbf{P}, \quad \mathbf{H} = \frac{1}{\mu_0}\mathbf{B} - \mathbf{M}, \tag{8.1.5}$$

$$\frac{d}{dt}\int_v \rho dv = 0, \tag{8.1.6}$$

$$\frac{d}{dt}\int_v \rho \mathbf{v} dv = \int_s \mathbf{t} ds + \int_v (\rho \mathbf{f} + \mathbf{F}^{EM}) = 0, \tag{8.1.7}$$

$$\frac{d}{dt}\int_v \mathbf{y} \times \rho \mathbf{v} dv = \int_s \mathbf{y} \times \mathbf{t} ds$$
$$+ \int_v [\mathbf{y} \times (\rho \mathbf{f} + \mathbf{F}^{EM}) + \mathbf{C}^{EM}] dv, \tag{8.1.8}$$

where $\mathbf{F}^{EM}$ is the electromagnetic body force and $\mathbf{C}^{EM}$ is the electromagnetic body couple. The energy equation in Eq. (5.1.22) needs to be generalized to include thermal effects in the following manner:

$$\frac{d}{dt}\int_v \rho \left(\frac{1}{2}\mathbf{v} \cdot \mathbf{v} + \varepsilon\right) dv = \int_v (\rho \mathbf{f} \cdot \mathbf{v} + \rho r) dv$$
$$+ \int_s (\mathbf{t} \cdot \mathbf{v} - \mathbf{n} \cdot \mathbf{q}) ds + \int_v W^{EM} dv, \tag{8.1.9}$$

where r is the body heat source per unit mass, $\mathbf{q}$ is the heat flux vector, and W^{EM} is the electromagnetic body power. In addition, the second law of thermodynamics needs to be added to the theoretical framework:

$$\frac{d}{dt}\int_v \rho \eta dv \geq \int_v \frac{\rho r}{\theta} dv - \int_s \frac{\mathbf{q} \cdot \mathbf{n}}{\theta} ds, \tag{8.1.10}$$

where θ is the absolute temperature and η is the entropy density per unit mass.

8.2   Differential Balance Laws

For differential balance laws, Eqs. (5.2.1)–(5.2.4), (5.2.7), (5.2.10) and (5.2.14) remain the same:

$$\nabla \times \mathbf{E} = -\frac{\partial \mathbf{B}}{\partial t}, \tag{8.2.1}$$

$$\nabla \times \mathbf{H} = \frac{\partial \mathbf{D}}{\partial t}, \tag{8.2.2}$$

$$\nabla \cdot \mathbf{D} = 0, \tag{8.2.3}$$

$$\nabla \cdot \mathbf{B} = 0, \tag{8.2.4}$$

$$\rho^0 = \rho J, \tag{8.2.5}$$

$$\rho \frac{d\mathbf{v}}{dt} = \nabla \cdot \boldsymbol{\tau} + \rho \mathbf{f} + \mathbf{F}^{EM}, \tag{8.2.6}$$

$$\varepsilon_{kij} \tau_{ij} + C_k^{EM} = 0. \tag{8.2.7}$$

The differential form of the energy equation takes the following form:

$$\rho \frac{d\varepsilon}{dt} = \tau_{ij} v_{j,i} + \rho E_i' \frac{d\pi_i}{dt} - M_i' \frac{dB_i}{dt} + \rho r - q_{i,i}. \tag{8.2.8}$$

The second law of thermodynamics is (see Eq. (3.12.4))

$$\rho \frac{d\eta}{dt} \geq \frac{\rho r}{\theta} - \left(\frac{q_i}{\theta} \right)_{,i}. \tag{8.2.9}$$

8.3 Nonlinear Constitutive Relations

For theoretical study it is usually desirable to have $\mathbf{E}'$, $\mathbf{B}$ and θ as independent constitutive variables. We introduce a free energy density per unit mass, F, through [20]

$$F = \varepsilon - E_i' \pi_i - \eta \theta. \tag{8.3.1}$$

Then the energy equation in Eq. (8.2.8) becomes:

$$\rho \frac{dF}{dt} + \rho \eta \frac{d\theta}{dt} + \rho \frac{d\eta}{dt} \theta$$

$$= \tau_{ij} v_{j,i} - P_i \frac{dE_i'}{dt} - M_i' \frac{dB_i}{dt} + \rho r - q_{i,i}. \tag{8.3.2}$$

The second law of thermodynamics in Eq. (8.2.9) can be written as

$$\rho r \leq \theta \left(\frac{q_i}{\theta} \right)_{,i} + \theta \rho \frac{d\eta}{dt} = \theta \left(\frac{q_{i,i}}{\theta} - \frac{q_i}{\theta^2} \theta_{,i} \right) + \theta \rho \frac{d\eta}{dt}$$

$$= q_{i,i} - \frac{q_i}{\theta} \theta_{,i} + \theta \rho \frac{d\eta}{dt}. \tag{8.3.3}$$

Eliminating r from Eqs. (8.3.2) and (8.3.3), we obtain the Clausius–Duhem inequality as

$$-\rho\left(\frac{dF}{dt} + \eta\frac{d\theta}{dt}\right) + \tau_{ij}v_{j,i} - P_i\frac{dE_i'}{dt} - M_i'\frac{dB_i}{dt} - \frac{q_i}{\theta}\theta_{,i} \geq 0. \quad (8.3.4)$$

We break $\boldsymbol{\tau}$, $\mathbf{P}$ and $\mathbf{M}'$ into reversible and dissipative parts as

$$\boldsymbol{\tau} = \boldsymbol{\tau}^R + \boldsymbol{\tau}^D,$$

$$\mathbf{P} = \mathbf{P}^R + \mathbf{P}^D, \quad (8.3.5)$$

$$\mathbf{M}' = \mathbf{M}'^R + \mathbf{M}'^D.$$

The reversible parts satisfy

$$\rho\frac{dF}{dt} + \rho\eta\frac{d\theta}{dt} = \tau_{ij}^R v_{j,i} - P_i^R\frac{dE_i'}{dt} - M_i'^R\frac{dB_i}{dt}. \quad (8.3.6)$$

Then the energy equation in Eq. (8.3.2) and the Clausius–Duhem inequality in Eq. (8.3.4) take the following form:

$$\rho\theta\frac{d\eta}{dt} = \tau_{ij}^D v_{j,i} - P_i^D\frac{dE_i'}{dt} - M_i'^D\frac{dB_i}{dt} + \rho r - q_{i,i}, \quad (8.3.7)$$

$$\tau_{ij}^D v_{j,i} - P_i^D\frac{dE_i'}{dt} - M_j'^D\frac{dB_j}{dt} - \frac{q_i}{\theta}\theta_{,i} \geq 0. \quad (8.3.8)$$

Equation (8.3.7) is the heat equation or dissipation equation.

With

$$v_{j,i} = X_{M,i}\frac{d}{dt}(y_{j,M}), \quad (8.3.9)$$

we write Eq. (8.3.6) as

$$\rho\frac{dF}{dt} = \tau_{ij}^R X_{M,i}\frac{d}{dt}(y_{j,M}) - P_i^R\frac{dE_i'}{dt}$$

$$- M_i'^R\frac{dB_i}{dt} - \rho\eta\frac{d\theta}{dt}. \quad (8.3.10)$$

Consider the following free energy:

$$F = F(y_{j,M}; E_i'; B_i; \theta). \qquad (8.3.11)$$

Then

$$\frac{dF}{dt} = \frac{\partial F}{\partial(y_{j,M})} \frac{d}{dt}(y_{j,M})$$

$$+ \frac{\partial F}{\partial E_i'} \frac{dE_i'}{dt} + \frac{\partial F}{\partial B_i} \frac{dB_i}{dt} + \frac{\partial F}{\partial \theta} \frac{d\theta}{dt}. \qquad (8.3.12)$$

Substituting Eq. (8.3.12) into Eq. (8.3.10), we obtain

$$\rho \frac{\partial F}{\partial(y_{j,M})} \frac{d}{dt}(y_{j,M}) + \rho \frac{\partial F}{\partial E_i'} \frac{dE_i'}{dt} + \rho \frac{\partial F}{\partial B_i} \frac{dB_i}{dt} + \rho \frac{\partial F}{\partial \theta} \frac{d\theta}{dt}$$

$$= \tau_{ij}^R X_{M,i} \frac{d}{dt}(y_{j,M}) - P_i^R \frac{dE_i'}{dt} - M_i'^R \frac{dB_i}{dt} - \rho\eta \frac{d\theta}{dt}, \qquad (8.3.13)$$

or

$$\left[X_{M,i}\tau_{ij}^R - \rho \frac{\partial F}{\partial(y_{j,M})} \right] \frac{d}{dt}(y_{j,M}) - \left(P_j^R + \rho \frac{\partial F}{\partial E_j'} \right) \frac{dE_j'}{dt}$$

$$- \left(M_j'^R + \rho \frac{\partial F}{\partial B_j} \right) \frac{dB_j}{dt} - \rho \left(\eta + \frac{\partial F}{\partial \theta} \right) \frac{d\theta}{dt} = 0. \qquad (8.3.14)$$

Equation (8.3.14) implies the following reversible constitutive relations:

$$\tau_{ij}^R = \rho y_{i,M} \frac{\partial F}{\partial(y_{j,M})}, \quad P_i^R = -\rho \frac{\partial F}{\partial E_i'}, \qquad (8.3.15)$$

$$M_i'^R = -\rho \frac{\partial F}{\partial B_i}, \quad \eta = -\frac{\partial F}{\partial \theta}. \qquad (8.3.16)$$

The dissipative parts of the constitutive relations, $\boldsymbol{\tau}^D$, $\mathbf{P}^D$, $\mathbf{M}'^D$ and $\mathbf{q}$, are restricted by Eq. (8.3.8).

For rotational invariance (objectivity), F can be reduced to a function of the following inner products and θ [20]:

$$C_{KL} = y_{i,K} y_{i,L},$$

$$\mathcal{E}_L = y_{i,L} E'_i, \quad B_L = y_{i,L} B_i. \tag{8.3.17}$$

We will use the strain tensor E_{KL} instead of the deformation tensor C_{KL}:

$$E_{KL} = (C_{KL} - \delta_{KL})/2. \tag{8.3.18}$$

Therefore, we take

$$F = F(E_{KL}; \mathcal{E}_K; B_K; \theta). \tag{8.3.19}$$

Then the constitutive relations in Eqs. (8.3.15) and (8.3.16) become

$$\tau_{ij}^R = \rho y_{i,M} \frac{\partial F}{\partial E_{ML}} y_{j,L}$$

$$+ \rho y_{i,M} \frac{\partial F}{\partial \mathcal{E}_M} E'_j + \rho y_{i,M} \frac{\partial F}{\partial B_M} B_j, \tag{8.3.20}$$

$$P_i^R = -\rho y_{i,L} \frac{\partial F}{\partial \mathcal{E}_L}, \tag{8.3.21}$$

$$M_i'^R = -\rho y_{i,L} \frac{\partial F}{\partial B_L}, \tag{8.3.22}$$

$$\eta = -\frac{\partial F}{\partial \theta}. \tag{8.3.23}$$

Alternatively, Eq. (8.3.6) can be written in material form as [18]:

$$\rho^0 \frac{dF}{dt} = P_{KL}^R \frac{dE_{KL}}{dt} - \mathcal{P}_K^R \frac{d\mathcal{E}_K}{dt}$$

$$- \mathcal{M}_K^R \frac{dB_K}{dt} - \rho^0 \eta \frac{d\theta}{dt}, \tag{8.3.24}$$

where

$$P_{KL}^R = JX_{K,i}X_{L,j}\tau_{ij}^{SR}, \quad \tau_{ij}^{SR} = \tau_{ij}^S - \tau_{ij}^{SD},$$
$$\mathcal{P}_K^R = JX_{K,i}P_i^R, \quad \mathcal{M}_K^R = JX_{K,i}M_i'^R. \tag{8.3.25}$$

Equation (8.3.24) suggests the following free energy directly:

$$F = F(E_{KL}; \mathcal{E}_K; B_K; \theta). \tag{8.3.26}$$

From Eqs. (8.3.24) and (8.3.26), we obtain the constitutive relations as

$$P_{KL}^R = \rho^0 \frac{\partial F}{\partial E_{KL}}, \quad \eta = -\frac{\partial F}{\partial \theta},$$
$$\mathcal{P}_K^R = -\rho^0 \frac{\partial F}{\partial \mathcal{E}_K}, \quad \mathcal{M}_K^R = -\rho^0 \frac{\partial F}{\partial B_K}, \tag{8.3.27}$$

which are equivalent to Eqs. (8.3.20)–(8.3.23).

With these constitutive relations, the angular momentum equation in Eq. (8.2.7) is satisfied automatically. The balance laws in Eqs. (8.2.1)–(8.2.6) and the dissipation equation in Eq. (8.3.7) can be written as equations for $\mathbf{y}$, $\mathbf{E}$, $\mathbf{B}$, ρ and θ. If $\mathbf{E}$ and $\mathbf{B}$ are expressed by the scalar and vector potentials through

$$\mathbf{B} = \nabla \times \mathbf{A},$$
$$\mathbf{E} = -\nabla\varphi - \frac{\partial \mathbf{A}}{\partial t}, \tag{8.3.28}$$

then the relevant balance laws can be written as equations for $\mathbf{y}$, φ, $\mathbf{A}$, ρ and θ.

8.4 Linear Materials

For linear electromagnetoelastic materials, we expand F into Taylor's series at

$$\theta = T_0, \quad E_{KL} = 0, \quad \mathcal{E}_K = 0, \quad B_K = 0. \tag{8.4.1}$$

We also denote the small temperature change from T_0 by

$$T = \theta - T_0, \quad |T| \ll T_0. \tag{8.4.2}$$

The constant and linear terms in the expansion of F are either immaterial or represent initial fields. Ignoring the constant and linear terms, we have the following expansion of F with second-order terms only [18]:

$$\rho^0 F = \frac{1}{2} c_{IJKL} E_{IJ} E_{KL} - \frac{1}{2} \chi^E_{KL} \mathcal{E}_K \mathcal{E}_L - \frac{1}{2} \chi^M_{KL} B_K B_L$$

$$- e_{KLM} \mathcal{E}_K E_{LM} - h_{KLM} B_K E_{LM} - m_{KL} \mathcal{E}_K B_L$$

$$- \frac{1}{2} \frac{C}{T_0} T^2 - \lambda_{KL} T E_{KL} - p_K T \mathcal{E}_K - q_K T B_K, \tag{8.4.3}$$

where C is related to the specific heat of the material. λ_{KL} are the thermoelastic constants, p_K the pyroelectric constants, and q_K the pyromagnetic constants. The corresponding linear constitutive relations are

$$P^R_{KL} = c_{KLMN} E_{MN} - e_{MKL} \mathcal{E}_M - h_{MKL} B_M - \lambda_{KL} T, \tag{8.4.4}$$

$$\mathcal{P}^R_K = \chi^E_{KL} \mathcal{E}_L + e_{KLM} E_{LM} + m_{KL} B_L + p_K T, \tag{8.4.5}$$

$$\mathcal{M}^R_K = \chi^M_{KL} B_L + h_{KLM} E_{LM} + m_{LK} \mathcal{E}_L + q_K T, \tag{8.4.6}$$

$$\rho^0 \eta = \frac{C}{T_0} T + \lambda_{KL} E_{KL} + p_K \mathcal{E}_K + q_K B_K. \tag{8.4.7}$$

8.5 Electrical Conduction

Electrical conduction is beyond the two-continuum model in Chapter 4. A three-continuum model [45] is needed to describe electrical conduction, with the third continuum being a free-charge continuum for electrical conduction. The basic laws of physics may be applied to each continuum with interactions. They can be combined to form the basic laws for the combined continuum of three. Most of the derivations are parallel to Chapters 4–6 and 8. We outline the differences caused by the electrical conduction below.

The sum of the lattice and bound charge densities, μ^l and μ^b, is no longer assumed to be zero. Instead, their sum is the lattice residual charge density μ^r:

$$\mu^l(\mathbf{y}, t) + \mu^b(\mathbf{y} + \boldsymbol{\eta}, t) = \mu^r(\mathbf{y}, t). \tag{8.5.1}$$

The residual charge is conserved, i.e.,

$$\frac{\partial \mu^r}{\partial t} + (\mu^r v_k)_{,k} = 0. \tag{8.5.2}$$

The free charge continuum is assumed to be massless. Its charge density is μ^e and its velocity is $\mathbf{v}^e$. We have

$$\frac{\partial \mu^e}{\partial t} + (\mu^e v_k^e)_{,k} = 0. \tag{8.5.3}$$

We denote the total charge and total current density by

$$\begin{aligned} \mu &= \mu^r + \mu^e, \\ \mathbf{J} &= \mu^r \mathbf{v} + \mu^e \mathbf{v}^e, \end{aligned} \tag{8.5.4}$$

which can be shown to satisfy

$$\frac{\partial \mu}{\partial t} + J_{k,k} = 0. \tag{8.5.5}$$

Two of the four Maxwell's equations are source free and remain the same as in Chapter 5. The other two with charges and currents are

$$\nabla \times \mathbf{H} = \frac{\partial \mathbf{D}}{\partial t} + \mathbf{J}, \tag{8.5.6}$$

$$\nabla \cdot \mathbf{D} = \mu. \tag{8.5.7}$$

The electromagnetic body force becomes

$$\begin{aligned} \mathbf{F}^{EM} &= \mathbf{P} \cdot \nabla \mathbf{E} + \mathbf{M}' \cdot (\mathbf{B}\nabla) \\ &\quad + \mathbf{v} \times (\mathbf{P} \cdot \nabla \mathbf{B}) + \rho \dot{\boldsymbol{\pi}} \times \mathbf{B} + \mu \mathbf{E} + \mathbf{J} \times \mathbf{B}. \end{aligned} \tag{8.5.8}$$

The electromagnetic body couple remains the same while the electromagnetic body power takes the following form:

$$\begin{aligned} W^{EM} &= (\mathbf{P} \cdot \nabla \mathbf{E}) \cdot \mathbf{v} + \rho \mathbf{E} \cdot \dot{\boldsymbol{\pi}} - \mathbf{M}' \cdot \frac{\partial \mathbf{B}}{\partial t} + \mathbf{E} \cdot \mathbf{J} \\ &= \mathbf{F}^{EM} \cdot \mathbf{v} + \rho \mathbf{E}' \cdot \dot{\boldsymbol{\pi}} - \mathbf{M}' \cdot \dot{\mathbf{B}} + \mathbf{J}' \cdot \mathbf{E}', \end{aligned} \tag{8.5.9}$$

where

$$\mathbf{J}' = \mathbf{J} - \mu\mathbf{v} = \mu^e(\mathbf{v}^e - \mathbf{v}). \qquad (8.5.10)$$

The energy equation becomes

$$\rho\frac{d\varepsilon}{dt} = \tau_{ij}v_{j,i} + \rho E_i'\frac{d\pi_i}{dt} - M_i'\frac{dB_i}{dt}$$
$$+ J_i'E_i' + \rho r - q_{i,i}. \qquad (8.5.11)$$

The reversible constitutive relations are still the same as those in Section 8.3. In addition, a constitutive relation for $\mathbf{J}'$ is needed which is restricted by the following Clausius–Duhem inequality:

$$\tau_{ij}^D v_{j,i} - P_i^D\frac{dE_i'}{dt} - M_j'^D\frac{dB_j}{dt} + J_i'E' - \frac{q_i}{\theta}\theta_{,i} \geq 0. \qquad (8.5.12)$$

Appendix A

List of Symbols

R_G – fixed reference frame (inertial)
R_C – instantaneous local rest frame (inertial)
$\mathbf{i}_1, \mathbf{i}_2, \mathbf{i}_3$ ($\mathbf{I}_1, \mathbf{I}_2, \mathbf{I}_3$) – basis vectors of Cartesian coordinates
$\mathbf{i}, \mathbf{j}, \mathbf{k}$ ($\mathbf{e}_1, \mathbf{e}_2, \mathbf{e}_3$) – basis vectors of Cartesian coordinates
$\mathbf{e}_r, \mathbf{e}_\theta, \mathbf{e}_z$ – basis vectors of cylindrical coordinates
i, j, k (I, J, K) – tensor indices. Range: 1–3
δ_{ij}, δ_{KL} – Kronecker delta
δ_{iK}, δ_{Ki} – coordinate transformation coefficients
$\varepsilon_{ijk}, \varepsilon_{IJK}$ – permutation symbol
q – elementary charge
ε_0 – vacuum electric permittivity
μ_0 – vacuum magnetic permeability
c, c_0 – light speed
k_B – Boltzmann constant
E_i – electric field in R_G
P_i – electric polarization per unit volume in R_G
D_i – electric displacement in R_G
B_i – magnetic flux or induction vector in R_G
H_i – magnetic field vector in R_G
M_i – magnetization per unit volume in R_G
S_i – Poynting vector

G_i – electromagnetic momentum density in $\mathrm{R_G}$

g_i – electromagnetic momentum density per unit mass

ρ^T – total charge per unit volume

ρ^P – polarization charge per unit volume

σ^P – polarization charge per unit area

ρ^E – difference of total and polarization charges $(\rho^T - \rho^P)$

σ^E – charge per unit area

p, n – concentrations of holes and electrons

γ^p, γ^n – body sources of holes and electrons

$\mathbf{J}^p, \mathbf{J}^n$ – current densities of holes and electrons

φ – electrostatic potential or scalar potential

$\mathbf{A}$ – vector potential

ψ – magnetostatic potential

$\mathbf{F}^E$ – electrostatic body force

$\mathbf{T}^E$ – electrostatic stress tensor

$\mathbf{C}^E$ – electrostatic body couple

$\mathbf{F}^M$ – magnetostatic body force

$\mathbf{T}^M$ – magnetostatic stress tensor

$\mathbf{C}^M$ – magnetostatic body couple

$\mathbf{F}^{EM}$ – electromagnetic body force

$\mathbf{T}^{EM}$ – electromagnetic stress tensor

$\mathbf{C}^{EM}$ – electromagnetic body couple

W^{EM} – electromagnetic body power

$\mathbf{J}^T$ – total current density

$\mathbf{J}^P$ – polarization current density

$\mathbf{J}^M$ – magnetization current density

$\mathbf{J}$ – difference of total and magnetization currents $(\mathbf{J}^T - \mathbf{J}^M)$

χ_{ij}^E – electric susceptibility

χ_{ij}^M, χ_{ij}^B – magnetic susceptibility

ε_{ij} – dielectric constants

μ_{ij} – magnetic permeability

U – internal energy per unit volume

H^E – electrostatic enthalpy per unit volume

H^M – magnetostatic enthalpy per unit volume

U^F – electromagnetic field energy per unit volume

$\widehat{U}$ – sum of field and internal energy densities $(= U^F + U)$

μ^l – lattice charge density

μ^b – bound charge density

μ^r – residual charge density $(\mu^l + \mu^b)$

μ^e – free charge density

E_i' – electric field in R_C

M_i' – magnetization in R_C

X_K – reference or material coordinates

y_i – present or spatial coordinates

u_K – mechanical displacement vector

J – Jacobian of deformation

C_{KL} – deformation tensor

E_{KL} – finite strain tensor

S_{kl} – linear strain tensor

v_i – velocity vector

a_i – acceleration vector

d_{ij} – deformation rate tensor

ω_{ij} – spin tensor

d/dt – material time derivative

$*$ – convective time derivative or complex conjugate

ρ – present mass density

ρ^0 – reference mass density

τ_{kl} – Cauchy stress tensor

K_{Lj} – first Piola–Kirchhoff stress tensor

P_{KL} – second Piola–Kirchhoff stress tensor

T_{kl} – linear stress tensor

ε – internal energy per unit mass

π_i – electric polarization per unit mass

χ – enthalpy per unit mass

F – free energy per unit mass

η – entropy per unit mass

θ – absolute temperature

q_i – heat flux vector

r – heat source per unit mass

p, q – matrix indices. Range: 1–6

c_{ijkl}, s_{ijkl} – elastic stiffness and compliance

e_{KIJ} – piezoelectric constants

h_{KIJ} − piezomagnetic constants

b^E_{KLMN} − electrostrictive constants

b^B_{KLMN} − magnetostrictive constants

m_{KL} − magnetoelectric constants

Appendix B

SI and Gaussian Units

SI	Gaussian
ε_0	1
μ_0	1
$c = 1/\sqrt{\varepsilon_0\mu_0} = 2.9979 \times 10^8 \text{m/s}$	$c = 2.9979 \times 10^{10} \text{cm/s}$
$\mathbf{E} = \dfrac{Q\mathbf{r}}{4\pi\varepsilon_0 r^3}$	$\mathbf{E} = \dfrac{Q\mathbf{r}}{r^3}$
$\mathbf{B} = \displaystyle\int \dfrac{\mu_0 \mathbf{J}(\mathbf{x}') \times \mathbf{r}}{4\pi r^3} dV'$	$\mathbf{B} = \displaystyle\int \dfrac{\mathbf{J}(\mathbf{x}') \times \mathbf{r}}{cr^3} dV'$
$\mathbf{F} = Q(\mathbf{E} + \mathbf{v} \times \mathbf{B})$	$\mathbf{F} = Q\left(\mathbf{E} + \dfrac{\mathbf{v}}{c} \times \mathbf{B}\right)$
$\nabla \times \mathbf{E} = -\dfrac{\partial \mathbf{B}}{\partial t}$	$\nabla \times \mathbf{E} = -\dfrac{1}{c}\dfrac{\partial \mathbf{B}}{\partial t}$
$\nabla \times \mathbf{H} = \dfrac{\partial \mathbf{D}}{\partial t} + \mathbf{J}$	$\nabla \times \mathbf{H} = \dfrac{1}{c}\dfrac{\partial \mathbf{D}}{\partial t} + \dfrac{4\pi}{c}\mathbf{J}$
$\nabla \cdot \mathbf{D} = \rho^E$	$\nabla \cdot \mathbf{D} = 4\pi\rho^E$
$\nabla \cdot \mathbf{B} = 0$	$\nabla \cdot \mathbf{B} = 0$

SI	Gaussian
$\mathbf{D} = \varepsilon_0\mathbf{E} + \mathbf{P}$	$\mathbf{D} = \mathbf{E} + 4\pi\mathbf{P}$
$\mathbf{B} = \mu_0\mathbf{H} + \mu_0\mathbf{M}$	$\mathbf{B} = \mathbf{H} + 4\pi\mathbf{M}$
$\mathbf{S} = \mathbf{E} \times \mathbf{H}$	$\mathbf{S} = \dfrac{c}{4\pi}\mathbf{E} \times \mathbf{H}$
$\mathbf{G} = \varepsilon_0\mathbf{E} \times \mathbf{B}$	$\mathbf{G} = \dfrac{c}{4\pi}\mathbf{E} \times \mathbf{B}$
$dw = \mathbf{E} \cdot d\mathbf{D} + \mathbf{H} \cdot d\mathbf{B}$	$dw = \dfrac{1}{4\pi}(\mathbf{E} \cdot d\mathbf{D} + \mathbf{H} \cdot d\mathbf{B})$

Appendix C

Vector Identities

For scalar fields f and g as well as vector fields $\mathbf{a}$, $\mathbf{b}$, $\mathbf{c}$ and $\mathbf{d}$,

$$\mathbf{a} \times (\mathbf{b} \times \mathbf{c}) = (\mathbf{a} \cdot \mathbf{c})\, \mathbf{b} - (\mathbf{a} \cdot \mathbf{b})\mathbf{c}$$

$$(\mathbf{a} \times \mathbf{b}) \cdot (\mathbf{c} \times \mathbf{d}) = (\mathbf{a} \cdot \mathbf{c})(\mathbf{b} \cdot \mathbf{d}) - (\mathbf{a} \cdot \mathbf{d})(\mathbf{b} \cdot \mathbf{c})$$

$$\nabla(fg) = g\nabla f + f\nabla g$$

$$\nabla \cdot (f\mathbf{a}) = (\nabla f) \cdot \mathbf{a} + f(\nabla \cdot \mathbf{a})$$

$$\nabla \times (f\mathbf{a}) = (\nabla f) \times \mathbf{a} + f(\nabla \times \mathbf{a})$$

$$\nabla \cdot (\mathbf{a} \times \mathbf{b}) = (\nabla \times \mathbf{a}) \cdot \mathbf{b} - \mathbf{a} \cdot (\nabla \times \mathbf{b})$$

$$\nabla \times (\mathbf{a} \times \mathbf{b}) = (\mathbf{b} \cdot \nabla)\mathbf{a} - (\mathbf{a} \cdot \nabla)\mathbf{b}$$
$$+ (\nabla \cdot \mathbf{b})\mathbf{a} - (\nabla \cdot \mathbf{a})\mathbf{b}$$

$$\nabla \times (\nabla \times \mathbf{a}) = \nabla(\nabla \cdot \mathbf{a}) - \nabla^2 \mathbf{a}.$$

For a closed curve C enclosing an area S with a unit normal $\mathbf{n}$, Stokes' theorem states that

$$\oint_C \mathbf{dl} \cdot \mathbf{G} = \int_S (\nabla \times \mathbf{G}) \cdot \mathbf{n}dS.$$

In addition [7],

$$\oint_C \mathbf{dl} \times \mathbf{G} = \int_S (\mathbf{n} \times \nabla) \times \mathbf{G} dS.$$

For a closed surface S with an outward unit normal $\mathbf{n}$ enclosing a volume V, the divergence theorem is

$$\int_S \mathbf{n} \cdot \mathbf{G} dS = \int_V \nabla \cdot \mathbf{G} dV.$$

Appendix D

Material Constants

Vacuum electric permittivity $\varepsilon_0 = 8.854 \times 10^{-12}$ F/m
Vacuum magnetic permeability $\mu_0 = 12.57 \times 10^{-7}$ H/m
Elementary charge $q = 1.602 \times 10^{-19}$ C
Boltzmann constant $k_B = 1.381 \times 10^{-23}$ J/K
$k_B T/q = 0.0259$ V at room temperature 300 K [46]
Planck's constant $\hbar = 1.05442 \times 10^{-27}$ erg-s

Aluminum nitride (AlN) [47]

$$\rho = 3{,}260 \, \text{kg/m}^3$$

$$[c_{pq}] = \begin{bmatrix} 345 & 125 & 120 & 0 & 0 & 0 \\ 125 & 345 & 120 & 0 & 0 & 0 \\ 120 & 120 & 395 & 0 & 0 & 0 \\ 0 & 0 & 0 & 118 & 0 & 0 \\ 0 & 0 & 0 & 0 & 118 & 0 \\ 0 & 0 & 0 & 0 & 0 & 110 \end{bmatrix} \times 10^9 \, \text{N/m}^2$$

$$[e_{ip}] = \begin{bmatrix} 0 & 0 & 0 & 0 & -0.48 & 0 \\ 0 & 0 & 0 & -0.48 & 0 & 0 \\ -0.58 & -0.58 & 1.55 & 0 & 0 & 0 \end{bmatrix} \, \text{C/m}^2$$

$$[\varepsilon_{ij}] = \begin{bmatrix} 8.0 & 0 & 0 \\ 0 & 8.0 & 0 \\ 0 & 0 & 9.5 \end{bmatrix} \times 10^{-11} \, \text{F/m}$$

Barium titanate (BaTiO$_3$) [48]

$$\rho = 5{,}800 \text{ kg/m}^3$$

$$[c_{pq}] = \begin{bmatrix} 166.0 & 77.0 & 78.0 & 0 & 0 & 0 \\ 77.0 & 166.0 & 78.0 & 0 & 0 & 0 \\ 78.0 & 78.0 & 162.0 & 0 & 0 & 0 \\ 0 & 0 & 0 & 43.0 & 0 & 0 \\ 0 & 0 & 0 & 0 & 43.0 & 0 \\ 0 & 0 & 0 & 0 & 0 & 44.5 \end{bmatrix} \times 10^9 \text{ N/m}^2$$

$$[e_{jq}] = \begin{bmatrix} 0 & 0 & 0 & 0 & 11.6 & 0 \\ 0 & 0 & 0 & 11.6 & 0 & 0 \\ -4.4 & -4.4 & 18.6 & 0 & 0 & 0 \end{bmatrix} \text{ C/m}^2$$

$$[\varepsilon_{ij}] = \begin{bmatrix} 11.2 & 0 & 0 \\ 0 & 11.2 & 0 \\ 0 & 0 & 12.6 \end{bmatrix} \times 10^{-9} \text{ F/m}$$

Cadmium selenide (CdSe) [49]

$$\rho = 4{,}820 \text{ kg/m}^3$$

$$[c_{pq}] = \begin{bmatrix} 90.7 & 58.1 & 51.0 & 0 & 0 & 0 \\ 58.1 & 90.7 & 51.0 & 0 & 0 & 0 \\ 51.0 & 51.0 & 93.8 & 0 & 0 & 0 \\ 0 & 0 & 0 & 15.04 & 0 & 0 \\ 0 & 0 & 0 & 0 & 15.04 & 0 \\ 0 & 0 & 0 & 0 & 0 & 16.3 \end{bmatrix} \times 10^9 \text{ N/m}^2$$

$$[e_{ip}] = \begin{bmatrix} 0 & 0 & 0 & 0 & -0.21 & 0 \\ 0 & 0 & 0 & -0.21 & 0 & 0 \\ -0.24 & -0.24 & 0.44 & 0 & 0 & 0 \end{bmatrix} \text{ C/m}^2$$

$$[\varepsilon_{ij}] = \begin{bmatrix} 9.02 & 0 & 0 \\ 0 & 9.02 & 0 \\ 0 & 0 & 9.53 \end{bmatrix} \varepsilon_0$$

Gallium arsenide (GaAs) [46,49,50]

$$\rho = 5{,}307\,\text{kg/m}^3$$

$$[c_{pq}] = \begin{bmatrix} 11.88 & 5.38 & 5.38 & 0 & 0 & 0 \\ 5.38 & 11.88 & 5.38 & 0 & 0 & 0 \\ 5.38 & 5.38 & 11.88 & 0 & 0 & 0 \\ 0 & 0 & 0 & 5.94 & 0 & 0 \\ 0 & 0 & 0 & 0 & 5.94 & 0 \\ 0 & 0 & 0 & 0 & 0 & 5.94 \end{bmatrix} \times 10^{10}\,\text{N/m}^2$$

$$[e_{ip}] = \begin{bmatrix} 0 & 0 & 0 & 0.154 & 0 & 0 \\ 0 & 0 & 0 & 0 & 0.154 & 0 \\ 0 & 0 & 0 & 0 & 0 & 0.154 \end{bmatrix} \text{C/m}^2$$

$$[\varepsilon_{ij}] = \begin{bmatrix} 12.5 & 0 & 0 \\ 0 & 12.5 & 0 \\ 0 & 0 & 12.5 \end{bmatrix} \times 8.85 \times 10^{-12}\,\text{F/m}$$

$$\mu^n = 8{,}500\,\text{cm}^2/\text{V}\cdot\text{s}$$
$$\mu^p = 400\,\text{cm}^2/\text{V}\cdot\text{s}$$

Germanium (Ge) [49,50]

$$\rho = 5{,}327\,\text{kg/m}^3$$

$$[c_{pq}] = \begin{bmatrix} 128.9 & 48.3 & 48.3 & 0 & 0 & 0 \\ 48.3 & 128.9 & 48.3 & 0 & 0 & 0 \\ 48.3 & 48.3 & 128.9 & 0 & 0 & 0 \\ 0 & 0 & 0 & 67.1 & 0 & 0 \\ 0 & 0 & 0 & 0 & 67.1 & 0 \\ 0 & 0 & 0 & 0 & 0 & 67.1 \end{bmatrix} \times 10^{9}\,\text{N/m}^2$$

$$[\varepsilon_{ij}] = \begin{bmatrix} 0.1398932 & 0 & 0 \\ 0 & 0.1398932 & 0 \\ 0 & 0 & 0.1398932 \end{bmatrix} \times 10^{-9}\,\text{F/m}$$

$$\mu^n = 3{,}900\,\text{cm}^2/\text{V}\cdot\text{s}$$
$$\mu^p = 1{,}900\,\text{cm}^2/\text{V}\cdot\text{s}$$

Lithium niobate (LiNbO₃) [51,52]

$$\rho = 4{,}700\,\text{kg/m}^3$$

$$[c_{pq}] = \begin{bmatrix} 2.03 & 0.53 & 0.75 & 0.09 & 0 & 0 \\ 0.53 & 2.03 & 0.75 & -0.09 & 0 & 0 \\ 0.75 & 0.75 & 2.45 & 0 & 0 & 0 \\ 0.09 & -0.09 & 0 & 0.60 & 0 & 0 \\ 0 & 0 & 0 & 0 & 0.60 & 0.09 \\ 0 & 0 & 0 & 0 & 0.09 & 0.75 \end{bmatrix} \times 10^{11}\,\text{N/m}^2$$

$$[e_{ip}] = \begin{bmatrix} 0 & 0 & 0 & 0 & 3.70 & -2.50 \\ -2.50 & 2.50 & 0 & 3.70 & 0 & 0 \\ 0.20 & 0.20 & 1.30 & 0 & 0 & 0 \end{bmatrix} \text{C/m}^2$$

$$[\varepsilon_{ij}] = \begin{bmatrix} 38.9 & 0 & 0 \\ 0 & 38.9 & 0 \\ 0 & 0 & 25.7 \end{bmatrix} \times 10^{-11}\,\text{F/m}$$

Lithium tantalate (LiTaO₃) [51,52]

$$\rho = 7{,}450\,\text{kg/m}^3$$

$$[c_{pq}] = \begin{bmatrix} 2.33 & 0.47 & 0.80 & -0.11 & 0 & 0 \\ 0.47 & 2.33 & 0.80 & 0.11 & 0 & 0 \\ 0.80 & 0.80 & 2.75 & 0 & 0 & 0 \\ -0.11 & -0.11 & 0 & 0.94 & 0 & 0 \\ 0 & 0 & 0 & 0 & 0.94 & -0.11 \\ 0 & 0 & 0 & 0 & -0.11 & 0.93 \end{bmatrix}$$
$$\times 10^{11}\,\text{N/m}^2$$

$$[e_{ip}] = \begin{bmatrix} 0 & 0 & 0 & 0 & 2.6 & -1.6 \\ -1.6 & 1.6 & 0 & 2.6 & 0 & 0 \\ 0 & 0 & 1.9 & 0 & 0 & 0 \end{bmatrix} \text{C/m}^2$$

$$[\varepsilon_{ij}] = \begin{bmatrix} 36.3 & 0 & 0 \\ 0 & 36.3 & 0 \\ 0 & 0 & 38.2 \end{bmatrix} \times 10^{-11}\,\text{F/m}$$

PZT-2 [49]

$$\rho = 7,600\,\mathrm{kg/m^3}$$

$$[c_{pq}] = \begin{bmatrix} 13.5 & 6.79 & 6.81 & 0 & 0 & 0 \\ 6.79 & 13.5 & 6.81 & 0 & 0 & 0 \\ 6.81 & 6.81 & 11.3 & 0 & 0 & 0 \\ 0 & 0 & 0 & 2.22 & 0 & 0 \\ 0 & 0 & 0 & 0 & 2.22 & 0 \\ 0 & 0 & 0 & 0 & 0 & 3.36 \end{bmatrix} \times 10^{10}\,\mathrm{N/m^2}$$

$$[e_{ip}] = \begin{bmatrix} 0 & 0 & 0 & 0 & 9.8 & 0 \\ 0 & 0 & 0 & 9.8 & 0 & 0 \\ -1.9 & -1.9 & 9.0 & 0 & 0 & 0 \end{bmatrix}\,\mathrm{C/m^2}$$

$$[\varepsilon_{ij}] = \begin{bmatrix} 504\varepsilon_0 & 0 & 0 \\ 0 & 504\varepsilon_0 & 0 \\ 0 & 0 & 260\varepsilon_0 \end{bmatrix}$$

PZT-4 [48]

$$\rho = 7,600\,\mathrm{kg/m^3}$$

$$[c_{pq}] = \begin{bmatrix} 138.5 & 77.37 & 73.64 & 0 & 0 & 0 \\ 77.37 & 138.5 & 73.64 & 0 & 0 & 0 \\ 73.64 & 73.64 & 114.8 & 0 & 0 & 0 \\ 0 & 0 & 0 & 25.6 & 0 & 0 \\ 0 & 0 & 0 & 0 & 25.6 & 0 \\ 0 & 0 & 0 & 0 & 0 & 30.6 \end{bmatrix} \times 10^{9}\,\mathrm{N/m^2}$$

$$[e_{jq}] = \begin{bmatrix} 0 & 0 & 0 & 0 & 12.72 & 0 \\ 0 & 0 & 0 & 12.72 & 0 & 0 \\ -5.2 & -5.2 & 15.08 & 0 & 0 & 0 \end{bmatrix}\,\mathrm{C/m^2}$$

$$[\varepsilon_{ij}] = \begin{bmatrix} 13.06 & 0 & 0 \\ 0 & 13.06 & 0 \\ 0 & 0 & 11.15 \end{bmatrix} \times 10^{-9}\,\mathrm{F/m}$$

PZT-5A [48]

$$\rho = 7{,}750\,\text{kg/m}^3$$

$$[c_{pq}] = \begin{bmatrix} 99.201 & 54.016 & 50.778 & 0 & 0 & 0 \\ 54.016 & 99.201 & 50.778 & 0 & 0 & 0 \\ 50.788 & 50.788 & 86.856 & 0 & 0 & 0 \\ 0 & 0 & 0 & 21.1 & 0 & 0 \\ 0 & 0 & 0 & 0 & 21.1 & 0 \\ 0 & 0 & 0 & 0 & 0 & 22.6 \end{bmatrix}$$

$$\times 10^9\,\text{N/m}^2$$

$$[e_{jq}] = \begin{bmatrix} 0 & 0 & 0 & 0 & 12.322 & 0 \\ 0 & 0 & 0 & 12.322 & 0 & 0 \\ -7.209 & -7.209 & 15.118 & 0 & 0 & 0 \end{bmatrix}\,\text{C/m}^2$$

$$[\varepsilon_{ij}] = \begin{bmatrix} 15.3 & 0 & 0 \\ 0 & 15.3 & 0 \\ 0 & 0 & 15.0 \end{bmatrix} \times 10^{-9}\,\text{F/m}$$

PZT-5H [49]

$$\rho = 7{,}500\,\text{kg/m}^3$$

$$[c_{pq}] = \begin{bmatrix} 12.6 & 7.95 & 8.41 & 0 & 0 & 0 \\ 7.95 & 12.6 & 8.41 & 0 & 0 & 0 \\ 8.41 & 8.41 & 11.7 & 0 & 0 & 0 \\ 0 & 0 & 0 & 2.30 & 0 & 0 \\ 0 & 0 & 0 & 0 & 2.30 & 0 \\ 0 & 0 & 0 & 0 & 0 & 2.33 \end{bmatrix} \times 10^{10}\,\text{N/m}^2$$

$$[e_{ip}] = \begin{bmatrix} 0 & 0 & 0 & 0 & 17.0 & 0 \\ 0 & 0 & 0 & 17.0 & 0 & 0 \\ -6.5 & -6.5 & 23.3 & 0 & 0 & 0 \end{bmatrix}\,\text{C/m}^2$$

$$[\varepsilon_{ij}] = \begin{bmatrix} 1{,}700\varepsilon_0 & 0 & 0 \\ 0 & 1{,}700\varepsilon_0 & 0 \\ 0 & 0 & 1{,}470\varepsilon_0 \end{bmatrix}$$

Silicon (Si) [49,50]

$$\rho = 2,332 \, \text{kg/m}^3$$

$$[c_{pq}] = \begin{bmatrix} 16.57 & 6.39 & 6.39 & 0 & 0 & 0 \\ 6.39 & 16.57 & 6.39 & 0 & 0 & 0 \\ 6.39 & 6.39 & 16.57 & 0 & 0 & 0 \\ 0 & 0 & 0 & 7.956 & 0 & 0 \\ 0 & 0 & 0 & 0 & 7.956 & 0 \\ 0 & 0 & 0 & 0 & 0 & 7.956 \end{bmatrix}$$
$$\times 10^{10} \, \text{N/m}^2$$

$$[\varepsilon_{ij}] = \begin{bmatrix} 11.7\varepsilon_0 & 0 & 0 \\ 0 & 11.7\varepsilon_0 & 0 \\ 0 & 0 & 11.7\varepsilon_0 \end{bmatrix}$$

$$\mu^n = 1,500 \, \text{cm}^2/\text{V} \cdot \text{s}$$
$$\mu^p = 450 \, \text{cm}^2/\text{V} \cdot \text{s}$$

Zinc oxide (ZnO) [49,50]

$$\rho = 5,680 \, \text{kg/m}^3$$

$$[c_{pq}] = \begin{bmatrix} 20.97 & 12.11 & 10.51 & 0 & 0 & 0 \\ 12.11 & 20.97 & 10.51 & 0 & 0 & 0 \\ 10.51 & 10.51 & 21.09 & 0 & 0 & 0 \\ 0 & 0 & 0 & 4.247 & 0 & 0 \\ 0 & 0 & 0 & 0 & 4.247 & 0 \\ 0 & 0 & 0 & 0 & 0 & 4.43 \end{bmatrix}$$
$$\times 10^{10} \, \text{N/m}^2$$

$$[e_{ip}] = \begin{bmatrix} 0 & 0 & 0 & 0 & -0.48 & 0 \\ 0 & 0 & 0 & -0.48 & 0 & 0 \\ -0.573 & -0.573 & 1.32 & 0 & 0 & 0 \end{bmatrix} \text{C/m}^2$$

$$[\varepsilon_{ij}] = \begin{bmatrix} 8.55\varepsilon_0 & 0 & 0 \\ 0 & 8.55\varepsilon_0 & 0 \\ 0 & 0 & 10.2\varepsilon_0 \end{bmatrix}$$

$$\mu^n = 200 \, \text{cm}^2/\text{V} \cdot \text{s}$$
$$\mu^p = 180 \, \text{cm}^2/\text{V} \cdot \text{s}$$

References

[1] S.H. Guo, *Electrodynamics*, Higher Education Press, Beijing, 1979 (in Chinese).

[2] A.T. Adams, *Electromagnetics for Engineers*, Ronald, New York, 1971.

[3] A.T. Adams and J.K. Lee, *Electromagnetics*, 2nd ed., Cognella, Solana Beach, CA, 2019.

[4] W.K.H. Panofsky and M. Phillips, *Classical Electricity and Magnetism*, Addison Willey, Reading, Massachusetts, 1962.

[5] L.D. Landau and E.M. Lifshitz, *Electrodynamics of Continuous Media*, 2nd ed., Butterworth–Heinemann, Linacre House, Jordan Hill, Oxford, 1984.

[6] H.F. Tiersten, *A Development of the Equations of Electromagnetism in Material Continua*, Springer, New York, 1990.

[7] G.E. Hay, *Vector and Tensor Analysis*, Dover, New York, 1953.

[8] C. Trimarco, A Lagrangian approach to electromagnetic bodies, *Tech. Mech.*, 22, 175–180, 2002.

[9] H.H. Woodson and J.R. Melcher, *Electromechanical Dynamics, Part I: Discrete Systems*, John Wiley & Sons, New York, 1968.

[10] A.C. Eringen, *Mechanics of Continua*, 2nd ed., Robert E. Krieger, Huntington, New York, 1980.

[11] I.S. Sokolnikoff, *Tensor Analysis*, 2nd ed., John Wiley and Sons, New York, 1964.

[12] H.F. Tiersten, Nonlinear electroelastic equations cubic in the small field variables, *J. Acoust. Soc. Am.*, 57, 660–666, 1975.

[13] X. Du, *Introduction to Continuum Mechanics*, Tsinghua University Press, Beijing, 1985 (in Chinese).

[14] R.D. Mindlin, *An Introduction to the Mathematical Theory of Vibrations of Elastic Plates*, J.S. Yang (ed.), World Scientific, Singapore, 2006.

[15] J.C. Baumhauer and H.F. Tiersten, Nonlinear electroelastic equations for small fields superposed on a bias, *J. Acoust. Soc. Am.*, 54, 1017–1034, 1973.

[16] H.F. Tiersten, On the accurate description of piezoelectric resonators subject to biasing deformations, *Int. J. Eng. Sci.*, 33, 2239–2259, 1995.

[17] H.F. Tiersten, On the nonlinear equations of thermoelectroelasticity, *Int. J. Engng. Sci.*, 9, 587–604, 1971.

[18] A.C. Eringen and G.A. Maugin, *Electrodynamics of Continua*, vol. I, Springer-Verlag, New York, 1990.

[19] D.F. Nelson, *Electric, Optic, and Acoustic Interactions in Dielectrics*, John Wiley and Sons, New York, 1979.

[20] H.F. Tiersten and C.F. Tsai, On the interaction of the electromagnetic field with heat conducting deformable insulators, *J. Math. Phys.*, 13, 361–378, 1972.

[21] H.G. de Lorenzi and H.F. Tiersten, On the interaction of the electromagnetic field with heat conducting deformable semiconductors, *J. Math. Phys.*, 16, 938–957, 1975.

[22] H.F. Tiersten, Coupled magnetomechanical equation for magnetically saturated insulators, *J. Math. Phys.*, 5, 1298–1318, 1964.

[23] J.J. Kyame, Wave propagation in piezoelectric crystals, *J. Acoust. Soc. Am.*, 21, 159–167, 1949.

[24] R.D. Mindlin, Electromagnetic radiation from a vibrating quartz plate, *Int. J. Solids Struct.*, 9, 697–702, 1972.

[25] P.C.Y. Lee, Electromagnetic radiation from an AT-cut quartz plate under lateral-field excitation, *J. Appl. Phys.*, 65, 1395–1399, 1989.

[26] J.S. Yang, A generalized variational principle for piezoelectromagnetism in an elastic medium, *Arch. Mech.*, 43, 795–798, 1991.

[27] J.S. Yang, Bleustein–Gulyaev waves in piezoelectromagnetic materials, *Int. J. Appl. Electromagn. Mech.*, 12, 235–240, 2000.

[28] J.L. Bleustein, A new surface wave in piezoelectric materials, *Appl. Phys. Lett.*, 13, 412–413, 1968.

[29] Y.V. Gulyaev, Electroacoustic surface waves in solids, *Sov. Phys. JETP Lett.*, 9, 37–38, 1969.

[30] J.S. Yang and H.G. Zhou, An interface wave in piezoelectromagnetic materials, *Int. J. Appl. Electromagn. Mech.*, 21, 63–68, 2005.

[31] C. Maerfeld and P. Tournois, Pure shear elastic waves guided by the interface of two semi-infinite media, *Appl. Phys. Lett.*, 19, 117–118, 1971.

[32] J.S. Yang, Acoustic gap waves in piezoelectromagnetic materials, *Math. Mech. Solids*, 11, 451–458, 2006.

[33] Y.V. Gulyaev and V.P. Plesskii, Acoustic gap waves in piezoelectric materials, *Sov. Phys. Acoust.*, 23, 410–413, 1977.

[34] D. Marcuse, *Light Transmission Optics*, Krieger, Florida, 1989.

[35] J.S. Yang, Piezoelectromagnetic waves in a ceramic plate, *IEEE Trans. Ultrason. Ferroelectr. Freq. Control*, 51, 1035–1039, 2004.

[36] J.L. Bleustein, Some simple modes of wave propagation in an infinite piezoelectric plate, *J. Acoust. Soc. Am.*, 45, 614–620, 1969.

[37] J.S. Yang, Love waves in piezoelectromagnetic materials, *Acta Mech.*, 168, 111–117, 2004.

[38] R.G. Curtis and M. Redwood, Transverse surface waves on a piezo-electric material carrying a metal layer of finite thickness, *J. Appl. Phys.*, 44, 2002–2007, 1973.

[39] S.N. Jiang, Q. Jiang, X.F. Li, S.H. Guo, H.G. Zhou and J.S. Yang, Piezoelectromagnetic waves in a ceramic plate between two ceramic half-spaces, *Int. J. Solids Struct.*, 43, 5799–5810, 2006.

[40] J.S. Yang, X.H. Chen and A-K Soh, Acoustic leakage in electromagnetic waveguides made from piezoelectric materials, *J. Appl. Phys.*, 101, 066105, 2007.

[41] J.S. Yang and S.H. Guo, Piezoelectromagnetic waves guided by the surface of a ceramic cylinder, *Acta Mech.*, 181, 199–205, 2006.

[42] C.L. Chen, On the electroacoustic waves guided by a cylindrical piezo-electric surface, *J. Appl. Phys.*, 44, 3841–3847, 1973.

[43] G.Y. Yang, J.K. Du, J. Wang and J.S. Yang, Frequency dependence of electromagnetic radiation from a finite vibrating piezoelectric body, *Mech. Res. Commun.*, 93, 163–168, 2018.

[44] J.S. Yang, A review of a few topics in piezoelectricity, *Appl. Mech. Rev.*, 59, 335–345, 2006.

[45] H.F. Tiersten, On the interaction of the electromagnetic field with deformable solid continua, in: *Electromagnetomechanical Interactions in Deformable Solids and Structures*, Y. Yamamoto and K. Miya, (eds.), North-Holland, 1987, pp. 277–284.

[46] R.F. Pierret, *Semiconductor Device Fundamentals*, Pearson, Uttar Pradesh, India, 1996.

[47] K. Tsubouchi, K. Sugai and N. Mikoshiba, AlN material constants evaluation and SAW properties on AlN/Al_2O_3 and AlN/Si, in: *Proc. IEEE Ultrasonics Symp.*, 1981, pp. 375–380.

[48] F. Ramirez, P.R. Heyliger and E. Pan, Free vibration response of two-dimensional magneto-electro-elastic laminated plates, *J. Sound Vib.*, 292, 626–644, 2006.

[49]   B.A. Auld, *Acoustic Fields and Waves in Solids*, vol. 1, Wiley, New York, 1973.

[50]   S.M. Sze, *Physics of Semiconductor Devices*, John Wiley & Sons, New York, 1981.

[51]   A.W. Warner, M. Onoe and G.A. Couqin, Determination of elastic and piezoelectric constants for crystals in class (3m), *J. Acoust. Soc. Am.*, 42, 1223–1231, 1967.

[52]   H.F. Tiersten, *Linear Piezoelectric Plate Vibrations*, Plenum, New York, 1969.

Index

Printed in the United States
by Baker & Taylor Publisher Services

Printed in the United States
by Baker & Taylor Publisher Services